Ivory Bridges

Ivory Bridges

Connecting Science and Society

Gerhard Sonnert
with the assistance of Gerald Holton

The MIT Press
Cambridge, Massachusetts
London, England

Set in Sabon by The MIT Press.

Library of Congress Cataloging-in-Publication Data

Sonnert, Gerhard, 1957–
Ivory bridges : connecting science and society / Gerhard Sonnert with the assistance of Gerald Holton.
p. cm.
Includes bibliographical references and index.
ISBN 978-0-262-19471-6 (hc. : alk. paper), 978-0-262-53692-9 (pb. : alk. paper)
1. Science—Social aspects. 2. Science and state. 3. Science—Societies, etc.
I. Holton, Gerald James. II. Title.
Q175.5 .S87 2002
303.48'3—dc21

2001056236

In memory of Henry W. Kendall, citizen-scientist (1926–1999)

Contents

Preface

The "ivory tower of science" is a widespread stereotype according to which science and society exist largely in isolation from each other. While this vision of an aloof science holds a good deal of truth, it is certainly not fully accurate. This book explores two "ivory bridges" that connect science with society. The first of these is federal science policy; the second consists of scientists' voluntary public-interest organizations. Because the public at large and even some parts of the scientific community appear to be rather unaware of the bridges, we intend to throw a strong light on these crucial links between science and society, both by presenting a scholarly study of these insufficiently attended to connections and by providing a substantial set of profiles of scientists' voluntary public-interest associations (appendix D).

Many people deserve our thanks for their help in making this publication possible. Our work on the Press-Carter Initiative in science policy benefited immensely from the generous input and cooperation of some major actors in that historical episode, first of all Frank Press himself. Donald Fredrickson, Robert Frosch, Gilbert Omenn, and Philip Smith graciously shared their recollections of the events. We also received valuable comments and suggestions from a number of other experts on science policy, among them William Blanpied, Lewis Branscomb, Harvey Brooks, Daryl Chubin, and Andrew Sessler. We are grateful to Frank Press for allowing us access to his materials archived at the Massachusetts Institute of Technology. We also thank the MIT Archives and the supportive staff there for facilitating our research.

Our study of scientists' public interest associations profited greatly from a meeting with a number of scientists, all knowledgeable in this area, who differed in their perspectives. Marcia Angell, Harvey Brooks, Paul Doty,

Carola Eisenberg, Stephen Jay Gould, John Holdren, Ruth Hubbard, Henry Kendall, and George Rathjens were at that gathering, and they all have our gratitude. In particular, we wish to acknowledge Henry Kendall, who only a few days after the meeting suffered a fatal accident. He was both a great scientist and a successful citizen-scientist.

Many other individuals also communicated valuable comments to us, among them Nina Byers, Kurt Gottfried, and Kosta Tsipis. Bill Williams generously provided information about British scientists' associations and drafted profiles of some of them. Furthermore, we thank the officers of scientists' associations who gave us feedback on our draft profiles of their organizations.

We were, once more, lucky to be able to benefit from Joan Laws's skill in administrative oversight. We also thank our student assistants, Kata Gellen and Tian Mayimin, who conscientiously helped us with the work on associations.

Finally, we gratefully acknowledge financial support from the William F. Milton Fund of Harvard University (for our study of the Press-Carter Initiative) and a Planning Grant from the American Academy of Arts and Sciences (for our study of scientists' public-interest associations). Neither of these institutions bears any responsibility for the contents of the book.

Ivory Bridges

1

From Ivory Tower to Ivory Bridges

A puzzling paradox motivates this book. On one hand, basic scientific research is, in most cases today, curiosity-driven; it is undertaken mainly in pursuit of intellectually challenging problems. The promise of socially useful by-products is not the foremost consideration, as one realizes quickly when glancing through the leading professional journals and when reading reports from laboratories or theoretical centers in astronomy, mathematics, the physical sciences, and much of biology. In addition, most scientific careers, certainly at the major research universities, flourish in proportion to the quantity and quality of intellectual triumphs, regardless of their applicability. For instance, one empirical study of how biologists evaluate one another (Sonnert 1995) showed intrinsic intellectual criteria of merit to be clearly predominant.[1] Moreover, the National Science Foundation has had great difficulty getting its peer reviewers to judge the broader societal impacts of research proposals. Most reviewers simply ignored a question about those impacts that in 1997 was included in the NSF's rating process (Mervis 2001).

At the extreme, some researchers may sympathize with the great mathematician G. H. Hardy, who boasted of being "the most unpractical member of the most unpractical profession of the world" and of having "never done anything 'useful'" (Hardy 1967: 43, 150). Even those who would not associate themselves consciously with that attitude are living and acting in, and are protected by, what is commonly called an ivory tower. Indeed, they would have been gravely hindered in their work without that protection, which has been constructed carefully for them over many years and which has made possible extraordinary feats of the imagination.

But on the other side of the paradox, we see the effects on society, for good or ill, that often result (occasionally with long time lags) from the mostly unforeseen spinoffs of basic research. It is almost a truism that science affects society more strongly now than ever before. Science has improved Americans' lives in countless ways; for example, it has helped people on average to live longer and healthier lives, and it has revolutionized our means of communication. Just as the steam engine became emblematic of the onset of industrial society in the nineteenth century, the electronic computer may come to symbolize our age, which Daniel Bell (1973) called "postindustrial."[2] Those developments have been amply discussed and need not be rehearsed again. Yet they lie at the heart of a paradox: at precisely the same time as the autonomy of science, proceeding according to its own internal logic, is great, so is its impact on society. Clearly, the current autonomy of science does not mean that there is no traffic between science and society—quite the contrary. This, then, marks our starting point.

What has been rather insufficiently attended to is an account of mechanisms that allow interchanges between basic science and society. The efforts (some successful, others stillborn) by scientists and other interested constituencies to build bridges between the ivory tower and the society beyond have been, and will remain, crucial both to the future of the scientific enterprise and to the health of society at large. Since World War II there have been two sustained political attempts to channel and control the societal ramifications of science. On the one hand, an increasingly elaborate system of governmental science policy has focused on creating the organizational preconditions for scientific progress as well as on harnessing such progress for political and societal benefits—both "policy for science" and "science in policy" (Brooks 1964: 76).[3] Our second chapter is devoted to this pathway. On the other hand, scientists have joined together in voluntary associations to oppose what they consider deleterious ramifications of science and technology for society, and to promote beneficent ones. We examine that pathway in our third chapter. (These two political connections are, of course, not the only ones. Additional links exist between science and society, notably a highly important economic bridge across which scientific knowledge is transferred via technology and product development to business and industry—a connection that has received extensive study.[4])

The topic of the two political infrastructures supporting the traffic between basic science and society—which so far have rarely been treated in conjunction—should be of interest to sociologists of science, to science policy makers and activists, and to scientists. Moreover, the very existence of the two bridges gives moral authority to the support and the conduct of basic research

The Rise of the Ivory Tower

The metaphor of the ivory tower originated in the nineteenth century. Ever since 1837, when Charles Augustin Sainte-Beuve was the first to record the term,[5] it has denoted the self-absorbed lifestyle of those who dedicate themselves single-mindedly to ethereal pursuits while ignoring the realities of life and society. Colleges and universities have been known as the primary building grounds of ivory towers, and many academic scientists have been prominent denizens of these structures.[6]

The intellectual roots and the legitimization of science for its own sake can be traced to the Humboldtian ideal of pure *Wissenschaft*, to which many leading American scientists and academics of the late nineteenth and the early twentieth century had been exposed during their training. Writing in 1937, Robert K. Merton (1968: 597) stated unequivocally that the concept of scientific autonomy was established clearly in most scientists' minds, and that it was "assimilated by the scientist from the very outset of his training."

In post-World War II American science, Vannevar Bush was highly influential in securing firm foundations for curiosity-driven research. He believed in protecting basic scientific research, not the least because, in his view, a high level of autonomy would allow basic science best to serve the societal good, largely through unpredictable spinoffs. Similarly, Bernard Barber (1953: 98) was convinced that "all research ultimately has some application," even though applications of pure research might take a long time to emerge. "However much 'pure' science may eventually be applied to some other social purpose than the construction of conceptual schemes for their own sake," Barber wrote (ibid.: 99), "its autonomy in whatever run of time is required for this latter purpose is the essential condition of any long run 'applied' effects it may have." In other words, because the benefits of pure research are not immediate, only the very autonomy of this research makes them possible. Don Price (1965: 177–178) appropriately observed that Vannevar Bush and others institutionalized after 1945 what the American scientists had already earlier come to believe: "Science has already freed itself philosophically from the idea of *ultimate* purpose;

American scientists at the end of the Second World War undertook to free the financing of basic research from subservience to political and economic purpose."

Buoyed by the institutional protections for curiosity-driven research put in place after World War II, most American scientists continued to follow a science-internal rationale of scientific work. Robert Morison (1979: xiii–xiv) portrayed it as follows: "Scientists on the whole have believed that the best, perhaps the only way to decide what to do next, is to look at the scientific work itself. Normally there is an inner logic to what is going on in any given laboratory as well as in the field as a whole." This devotion to science-internal imperatives was poetically captured by Stephen Toulmin (1975: 118): "Even today, most youthful apprentices to the scientific professions still adopt the same single-valued, single-minded rule of life as the religious novices of earlier ages. They devote themselves to their pursuit of better scientific understanding in a chosen discipline, with the same total commitment that the young medieval monk gave to his pursuit of Sanctity." Any deviation from this detached ideal of science is likely to provoke fierce opposition from scientists. "In the eyes of some scientists," Jean-Jacques Salomon (then head of the Science Policy Division of OECD) noted in 1973, "this change [i.e., science's becoming an integral part of the production system] seems like a betrayal of the ends (and the interests) of science, in so far as it should be concerned solely with the pursuit of truth. The horizon of utility under which it has blossomed compromises it, alienates it and, in a word, prostitutes it." (Salomon 1973: 173–174; see also Salomon 2001)

Critical Voices

The concept of science for science's sake, so cherished by most scientists, has not gone unchallenged. Our short survey of voices that oppose ivory-tower science starts with an extreme counter position held by a small but vociferous group of thinkers who reject all science wholesale. These individuals, often described as Romantic, are deeply hostile to what they consider a cold and fragmenting worldview promoted by science. In their opinion, science disenchants the mystic majesty of nature, subjects people to an alienated, machine-like existence, and, in effect, destroys their souls. The following pre-World War I lament by the German philosopher Ludwig Klages eloquently expresses this deep-felt anguish about the modern scientific-technological *Zivilisation*: "We were not mistaken when we suspected 'progress' of a vain desire for power, and we see that there is a method to the madness of destruction. Behind the pretexts of 'utility,' 'economic development,' and 'cultivation,' the real purpose [of 'progress'] is the *annihilation of life*. . . . No doubt, we live in the age of the *destruction of the soul*." (Klages 1956 [1913]: 12–14)

The passages quoted above sound less antiquated than their age might suggest because many contemporary intellectuals profess very similar views.[7] In 1992, for example, at the World Economic Forum in Davos, Switzerland, the distinguished Czech writer and statesman Václav Havel gave a speech titled "The End of the Modern Era" in which he severely criticized modernity (and communism as its primary culprit and extreme expression). "This [modern] era," Havel exclaimed, "has created the first global, or planetary, technical civilization, but it has reached the limit of its potential, the point beyond which the abyss begins. . . . Traditional science, with its usual coolness, can describe the different ways we might destroy ourselves, but it cannot offer us truly effective and practicable instructions on how to avert them." (*New York Times*, March 1, 1992)

A second (and internally varied) strand of critique is an epistemological relativism held by some sociologists and philosophers of science, including Harry Collins (1981) and Bruno Latour (1987).[8] In this view, science is considered just one particular belief system, one particular way of constructing the world, among many. A major intention of those who hold this view is to debunk science's customary claim to objective knowledge about the world. According to Collins (1981: 3), "one school, . . . inspired in particular by Wittgenstein and more lately by the phenomenologists and ethnomethodologists, embraces an explicit relativism in which the natural world has a small or non-existent role in the construction of scientific knowledge." Within a relativist framework, science may not necessarily be evil, as the Romantics would have it, but it hardly deserves an ivory tower or any special treatment.

Other challenges of the ivory tower have been supported also within the natural science community. Championed mainly by left-leaning thinkers, the idea that science must immediately and exclusively serve societal purposes has at times garnered a considerable following among scientists. That maxim reached an early peak of popularity after World War I, when the newly established Soviet Union expressly fashioned its science policy according to it.

The notion also attracted support among distinguished Western scholars and scientists, notably J. D. Bernal, J. B. S. Haldane, and P. M. S. Blackett.[9]

Greater social relevance also was advocated by some scientists who did not wish to abolish curiosity-driven research completely. For example, Bertrand Russell, one of the prototypes of the activist intellectual, criticized the isolationism prevalent among many scientists and hinted at the sinister real-life foundations of the lofty tower: "As the world becomes more technically unified, life in an ivory tower becomes increasingly impossible. Not only so; the man who stands out against the powerful organizations which control most of human activity is apt to find himself no longer in the ivory tower, with a wide outlook over a sunny landscape, but in the dark subterranean dungeon upon which the ivory tower was erected." (Russell 1960: 392)

Another notable voice in this tradition is the biophysicist Joseph Rotblat, who received the 1995 Nobel Peace Prize for his pioneering work in the Pugwash Conference movement. In his speech accepting the prize, Rotblat told the scientific community: "You are doing fundamental work, pushing forward the frontiers of knowledge, but often you do it without giving much thought to the impact of your work on society. Precepts such as 'science is neutral' or 'science has nothing to do with politics' still prevail. They are remnants of the ivory tower mentality, although the ivory tower was finally demolished by the Hiroshima bomb."[10]

Science in the United States: Dissent and Establishment

Criticisms of ivory-tower science can also be found beyond the small circles of intellectuals and scientists. In the United States, both criticism of science and respect for it are linked to deep-rooted cultural attitudes.[11]

Since the early days of European settlement, the American self-image had included an explicit and self-conscious component of a largely moralistic exceptionalism that set the new American project apart from what was perceived as the pervasive corruption of the Old World. Thomas Paine, in the revolutionary age, called on the new country "to begin the world over again." The founders of the United States were fully aware of the experimental nature of their act of creating a constitution unlike that of any other sovereign state. They felt a positive affinity toward science because they thought that science could provide support for this unique experiment.

Yaron Ezrahi (1990) and I. B. Cohen (1995) recount how the proponents of liberal democracy used scientific principles as cultural resources to legitimize their ideas of rational government and public action.

Another and perhaps even more powerful strain of American politics has been sympathetic to a particular kind of authority-defying scientists. Since the earliest colonial times, there has been distrust of any kind of "establishment" (that is, an institution which society is obliged to support but which itself is not responsible to society). The original anti-establishment impulses were directed against the Church of England and the British government, but in time they were transferred to American politics and redirected against the indigenous government.[12]

Science and religion are most often regarded as antithetical to each other, yet in some aspects scientists have resembled the early religious dissenters who came to settle in the New World (Price 1985). Science has been characterized as "the structured and sanctioned overthrowing of authority" (Ausubel 1998: 9). In keeping with Bronowski's apt observation (1965: 61) that "dissent is the native activity of the scientist," a pronounced iconoclastic streak can be detected in many scientists. They tend to revel in challenging accepted knowledge with new data or new theories, and they can often be found embarking on a quest for knowledge with relatively little regard for political or social expediency. (The scientists' truth is, of course, based on empirical evidence, whereas the religious dissenters' truth rests on religious authority or on personal supernatural experience.) In America's formative age, these traits of independent-mindedness and dissent secured scientists a place among the primordial American icons. (They were joined, during the westward expansion, by trappers, gold miners, and cowboys, and other popular icons of hardy frontier folk, all self-reliant and not particularly obsequious to any kind of authority.) When Joseph Priestley—equally unorthodox as a chemist, a political writer, and a theologian—arrived in the United States, having fled England, where a mob had destroyed his house, his library, and his laboratory, Thomas Jefferson embraced him as a fellow dissident.

Scientific research tends to be more expensive than religious dissent. Whereas it has been feasible (controversies about details notwithstanding) to separate state and religion and to leave individuals to their own religious beliefs and practices without government interference, state and science have moved ever closer together. Since the middle of the twentieth century, the US government has provided massive financial support of science and has introduced scientists into various governmental offices and functions, especially in the executive branch.[13] Because scientists (collectively) have joined the long line of societal constituencies who appear in Washington, asking for support, "most members of Congress think of scientists as another interest group. Perhaps smarter lobbyists . . . the science lobby. They are coming in asking for more research and development money . . . more instruments, better telescopes."[14] Yet, while requesting public support, scientists as a group still assert a high degree of autonomy from political control. They are an elite, and elites in are usually viewed with some suspicion in the United States. Consequently, science could be, and indeed often is, distrusted as being a modern establishment. Many politicians are wary of handing the scientific community money "with no strings attached." In a way, the motto "Science on tap, but not on top," popular among politicians and bureaucrats, may hint at their fears of losing control over a costly science system without clearly visible societal benefits, and of thus creating a new kind of establishment. The scientists' current collective social position has been perceptively described as that of established dissenters (Price 1965)—a term of appropriate oxymoronic flavor, because this position certainly has also created tensions and potential conflicts for scientists and exposed their ivory tower to popular and political scrutiny.

Building Bridges

So far we have encountered scientists who love the splendid isolation of their ivory tower and an assortment of voices that criticize or even condemn it. But there has been a third way. Scientists and others have worked to keep the ivory tower intact but at the same time to reduce its isolation. With some success, they have attempted to build bridges to connect the ivory tower with society—bridges that might explain the initial paradox of the simultaneous boom of purely curiosity-driven research and of the societal impacts of science.

The scientists who are the focus of this book are often described as either "insiders" or "outsiders" (Perl 1971; Price 1965; Primack and von Hippel 1974). Following Don Price (1962), we shall call the insiders *scientist-administrators*[15]; following Frank von Hippel (1991), we shall call the outsiders *citizen-scientists*.[16] Whereas scientists in their (often temporary) roles as scientist-administrators work within the official structure of government (as science advisers, as officials at the Office of Science and Technology Policy or at the National Science Foundation, or as employees of a multitude of other government agencies), the citizen-scientists have come together in voluntary associations (such as the Federation of American Scientists and the Union of Concerned Scientists) to express views on science and science policy that are often opposed to specific or fundamental aspects of the government's stance. This book examines how each of these two groups—scientist-administrators and citizen-scientists—has worked to bridge the gap between the ivory tower and real life.

Each of the bridges is related to a hypothesis set forth by Alexis de Tocqueville in *Democracy in America*. Tocqueville's assertion that Americans look to science for its applications underlies our discussion of Jeffersonian Science in chapter 2, where we focus on the government's science policy apparatus; his assertion that Americans tend to form civil associations informs our discussion of scientists' organizations in chapter 3, where we ask how and why scientists band together to influence the public science agenda from the "outside."

A substantial body of first-rate scholarship already exists in the field of federal science policy.[17] Thus, rather than produce a wide-ranging overview that would repeat much that has already been said, we shall present a case study of an episode that, although relatively obscure, holds important lessons for present and future science policy. Since scientists' voluntary public-interest organizations have been severely neglected as a field of study, a general survey of them will be beneficial.

Jeffersonian Science

Chapter 2 of this book is about science policy, the government's support of research, and its use of scientific expertise. It has two main purposes. The first is historical: to chronicle the Press-Carter Initiative, a momentous yet now largely forgotten episode of the late 1970s. The second is forward-looking: to suggest the explicit and conscious adoption of what we shall define and describe as *Jeffersonian research*[18] as one of the rationales for future science policy.

The initiative by Frank Press and Jimmy Carter serves as a pertinent example of a science policy within a Jeffersonian framework. We propose that the notion of research in the Jeffersonian mode can and should advance the current debates about how the relationship between science and its federal supporting agencies should be redefined and about what kinds of research the federal government should support. An unfortunate dichotomy appears to hamper this debate: the popular dichotomy between basic science, undertaken without regard for its social applications, and societally useful applied science. Our notion of Jeffersonian research may help to overcome this false dichotomy by demonstrating how basic research can be closely coupled to the societal interest, and may thus lead to an improved science policy that maximizes the societal good that flows from basic research (Branscomb 1995, 1998, 1999; Branscomb, Holton, and Sonnert 2001; Brooks 1980, 1994; Holton 1986, 1998, 2001; Holton and Sonnert 1999; Sonnert and Brooks 2001).

In Tocqueville's day, a familiar argument contended that science could not flourish in a democracy. Individuals of scientific genius, it was asserted, would receive patronage from aristocratic rulers, who would respect grandeur of every sort, whereas they would be stifled under the rule of intellectual mediocrity typical in a democracy. Tocqueville opened his discussion of American science by rebutting this argument, but he did say that science in a democracy characteristically slants toward practical applications. In Tocqueville's view, the populace supports a science that delivers (or promises to deliver) practical benefits, but is much less eager to support science for its own sake—that is, science solely for the advancement of knowledge.

Galileo's explanation of why he left a position in the Republic of Venice for the patronage of the authoritarian Grand Duke of Tuscany might be seen as a forerunner of Tocqueville's hypothesis: "It is impossible to obtain wages from a republic, however splendid and generous it may be, without having duties attached. For to have anything from the public one must satisfy the public and not any one individual; and so long as I am capable of lecturing and serving, no one in the republic can exempt me from duty while I receive pay. In brief, I can hope to enjoy these benefits only from an absolute ruler." (Galileo 1957: 65) (From a modern perspective, one might add the caveat that some absolute rulers, far from being disinterested patrons of pure science, have been known to misuse science for their authoritarian ends.) In the democratic American framework of dissent and establishment, the American public may cheerfully tolerate maverick scientists' sometimes "weird" and apparently useless projects until it is asked to pay for them.

At first sight, Tocqueville's hypothesis looks plainly obsolete. For many decades, Americans have witnessed an amazing flowering of leading-edge basic research in their country. Yet one should not forget that it was the practical accomplishments of science and engineering during World War II that widely legitimized the turn to basic research. Decisive scientific-technological contributions to the war effort, such as radar and synthetic rubber, allowed Vannevar Bush and other scientific leaders of the time to surround basic research with the halo of these achievements even before the end of the war, and thus to call forth public support for it. The scientists' promise of further useful developments, backed by an imposing track record, as well as the people's fear of losing ground against rival scientific efforts in the Eastern Bloc, caused public funds to flow into basic research. No amount of brilliant breakthroughs in basic science may have been sufficient to secure that support. What Daniel Kevles (1987: ix) said about the physicists was true, by extension, of the whole scientific community: "To a considerable extent, what brought them to power is what keeps them there today—the identification of physics with the material and military elements of national security." With the basic parameters of this "contract" between science and society shifting since the end of the Cold War, it has again become a matter of intense debate why and to what extent government should support basic research, just as Tocqueville would have predicted.[19]

Once small and informal, science advising and science policy in the federal government expanded greatly by the late 1970s, when the events featured in the case study of chapter 2 occurred.[20] In the descriptive sections of chapter 2, we will focus on the complex dealings within the government that brought about the Press-Carter Initiative; we will not explicitly discuss the characteristics of the scientist-administrators in a general way. Therefore, we will briefly address that issue now.

Two views exist about the federal government's science policy structure. Simply stated, the view first regards scientist-administrators as a closed and distinct elite that is well versed in the Washington language of political power. (See, e.g., Perl 1971.) The other view regards them as a rotating non-elite group, fairly indistinguishable from the population of American scientists as a whole, that remains alienated from the power-driven culture of politics. (See, e.g., Mullins 1981.) These two statements are not as contradictory as they might seem, and they can easily be reconciled. They highlight different aspects of an internally differentiated and complex science policy structure that contains both an elite and a larger group of less distinguished affiliates, and both full-time administrators and part-time advisers. The elite is typically more acculturated than the rotating rank-and-file membership of the scientist-administrator group.

Yet even the most senior scientist-administrators rarely get acculturated completely into the Washington power system, as they owe allegiance to two cultures—the culture of science and the culture of power—that differ markedly. Scientist-administrators often baffle or irritate Washingtonians when they avoid broad speculation and stick to stringent standards of empirical evidence. Their advice brims with cautions, hedges, and alternative possibilities, whereas the action-oriented politicians and bureaucrats hope for fast and simple answers (Mullins 1981: 13). The Washingtonians' frustration with the scientists' usual advice is neatly captured in the proverbial cry "Give me a one-handed scientist!"—that is, one who cannot say "on the other hand." Politicians are also quicker to accept the influence of powerful interests on decisions and quicker to compromise. Political discourse emphasizing "power" differs from scientific discourse emphasizing "truth." Yet one should not overlook the fact that scientific debates can give rise to their own kind of acrimony. According to the inner logic of science, scientific issues cannot be decided by majority vote, and there is little room for compromises of the "split the difference" kind that often resolve political struggles.

The Nixon years—a particularly dramatic era in the history of science advising—may offer a good illustration of the problems of acculturation (Trenn 1983). Primack and von Hippel (1974: 20–21) recount how President Nixon's science adviser, Lee DuBridge, who initially had criticized the supersonic transport (SST) project, came around to defending the administration's approving stance. DuBridge is quoted as telling Representative Sidney Yates (D-Illinois): "Congressman, I am a soldier. The President has made up his mind, and I am going to support the President's decision." (ibid.: 21) Primack and von Hippel (ibid.) call this an "undignified posture." But why? Every day, countless federal bureaucrats are overruled by their superiors. Only rarely does one of them resign in indignation or try to take a defeated scheme to the public. Yet the opinion expressed by Primack and von Hippel—probably shared by a large part of the scientific community—is that scientist-administrators should retain a certain allegiance to the science-internal concern for truth and, more precisely, to dissenting in the name of truth. Subscribing to that tradition, Primack and von Hippel argue that it is the duty of the government's science advisers to dissent.

By becoming a self-described soldier of the administration, the scientist-administrator DuBridge was seen by his critics as having somehow betrayed his role as a scientist and turned into a mere administrator. Indeed, other members of the President's Science Advisory Committee (PSAC) were less pliable. The fact that a member of the PSAC publicly opposed the administration on the SST issue probably precipitated Nixon's decision in 1973 to abolish the top-level presidential science advisory system. Irritation had been mounting in the government since it had been noticed that some high-level scientist-administrators were not as functional as the regular policy administrators in Washington. An unnamed White House staffer had asked "Who in the hell do those science bastards think they are?" (Herken 1992: 180) This episode gave some credence to a central criticism of the official advisory system: that it provides merely "a facade of prestige which tends to legitimize all technical decisions made by the President" (Perl 1971: 1214). According to this somber view, scientific advice is appreciated as long as it supports what the government wants to do anyway. In that case, the advisers provide the welcome "scientific facts" to make government policy look reasonable. Otherwise, the advice is swept under the carpet.

Perl's discussion was published in 1971, shortly before the top layers of the official advisory structure were abolished and when the wave of activism from the outside was just gathering momentum (as Perl notes in passing). Perl did not believe in the effectiveness of working from outside the government. He reckoned that most leading scientists were on the inside, and that they were the influential role models for the scientific community. He was in favor of strengthening the element of dissent within the advisory structure. But then came the downfall of the official advisory structure. In 1974, Primack and von Hippel published their opposite conclusion: that it was more effective to work from the outside than from the inside. The intense outside efforts of the early 1970s were one burst of activity in a longer history of scientists' organizing themselves.

Scientists' Public-Interest Associations

In chapter 3 we shall examine the voluntary public-interest organizations that scientists have formed outside the government, often in opposition to government policies. These organizations correspond to Tocqueville's observation of the peculiarly American propensity for organization, which he found to pervade American society.

The issues of science policy are certainly not simple and straightforward. There have been heated disagreements about the risks and benefits of particular innovations; there have been equity concerns about who reaps the benefits, who bears the costs of scientific progress, and who controls it; and there have been conflicts between scientific findings and established economic, social, political, or cultural interests. Because this is an area of both great importance and great controversy, it should not be entirely surprising that the government's science policy apparatus has been complemented and even challenged by scientists' organizations.

In the activist-scientists' organizations, the old icon of the "scientist as dissenter" reasserts itself. It is important to note that the construction of this particular connection between science and society was spearheaded by members of the scientific community and allied public-interest activists, not by the public at large. "There are," Culliton (1979: 149) observes, "no convincing data that the general public is anti-science or that it wishes to undermine the scientific tradition. What is happening is something else; the scientists themselves are challenging the way the research enterprise is run."

The Tocquevillian postulate of Americans' proclivity for organizing has sometimes been used to contrast civic organizations and the government bureaucracy in an either-or fashion. As soon as government organizes more and more spheres of social life, this argument goes, civic organizations will dwindle—or, conversely, organizations will flourish if the government's interference in society is at a low level. In a study of large civic associations, Skocpol (1997: 467) argued that government activity and citizens' activity are not simply substitutes for one another and emphasized the "non-zero-sum nature of US governmental and associational expansion."

As we will show, the history of scientists' public-interest associations clearly supports Skocpol's more general point. Many of these groups arose in response to government policy, and many of them identify the government as the target of their activities and try to gain some control over the domain of the scientist-administrators. In other words, although our two bridges from the ivory tower are certainly rival pathways from science to society, they are historically not alternatives. On the contrary, the second bridge (that of the citizen-scientists) materialized largely *because* the first one was built. "Inside" and "outside" are related, and there are occasional crossovers between the scientist-administrators and the citizen-scientists. If scientists' societal goals and interests are not satisfied within the government (through involvement in the official science advisory structure), they might be expressed in organization outside the government. Consider this remark regarding the negotiations about the National Institutes of Health guidelines on recombinant DNA (Nelkin 1979: 198): "If dissenters cannot win their battles at one level, they typically seek to broaden their constituency. Thus [scientists opposing the NIH guidelines] sought a wider framework for negotiation through greater public involvement."

Analogous to the role conflict experienced by scientist-administrators, a potential "wearing-two-hats problem" also exists for the citizen-scientists. "Scientists are citizens too, and ought to be encouraged to participate fully in politics, to which they may make a unique contribution as long as they make clear the limits of their competence as scientists to answer unscientific questions." (Price 1979: 88–89) Just as scientist-administrators often feel pulled in different directions by the loyalty to an administration and the urge to dissent in the name of scientific integrity, citizen-scientists may experience an equally virulent conflict between the ethos of science that emphasizes objectivity and their political role as citizens who take action based on commitments to morals and values. As we shall discuss in detail in chapter 3, different scientists' organizations have taken different stances on this fundamental issue of where to draw the line between science and politics.

Theoretical Considerations: Systems and Boundaries

To get a deeper understanding of the issues, it may be useful to turn to the sociological theory of societal systems. Emile Durkheim, one of the founders of sociology, famously contrasted the relatively simple division of labor in traditional societies with the much more intricate division of labor in modern societies. Following Durkheim and other sociological pioneers, most present-day sociologists agree that a major hallmark of modern societies is a high degree of differentiation and complexity. Modern societies have developed a number of increasingly autonomous subsystems with semi-permeable boundaries. Niklas Luhmann (1990), one of the exponents of the functionalist analysis of society, emphasized that each of the emerging subsystems of modern society tends to develop its own rationale, an autonomous inner logic that sets it apart from the other subsystems. The history of science certainly exemplifies this general movement toward differentiation. Joseph Ben-David (1984: 169) highlighted the astounding transformation of modern science from its beginnings in the eighteenth century into "a relatively autonomous subsystem of society." In that process, science also has developed its own ethos (Merton 1973). Truth can be identified as a guiding principle of this scientific ethos or rationale (Bronowski 1965: 60; Luhmann 1990).[21] Price (1965: 183), who talked of estates rather than subsystems, similarly noted that the scientific and the political estate differed in their primary concerns: truth for science, purpose (or power) for politics.[22] Just as systems theorists would expect, the science subsystem has grown and differentiated from society, and this process has supported the detachment of scientific activity from extraneous rationales—that is, the "ivory tower" autonomy of modern science.

This theoretical perspective also directs attention to boundary management, a fertile research site. How do science and the other subsystems relate to one another, and how do interactions occur across their boundaries? An important strand of current scholarship about science and society has centered on how these boundaries are being negotiated (Gieryn 1983, 1999; Guston 2000a,b). In this sense, as we shall observe in chapter 3, the demarcation of politics and scientific expertise indeed has been a fundamental quandary for citizen-scientists.

2

A Jeffersonian Mode of Science Policy: The Press-Carter Initiative

The public attitude toward science is still largely positive in the United States.[1] For a vocal minority, however, the fear of risks and even catastrophes supposedly resulting from scientific progress has created a starkly negative view of science. Additionally, central claims of scientific epistemology have come under attack from academics outside science (e.g., Collins 1981; Latour 1987; compare Bricmont and Sokal 2001). Some portions of the political sector consider basic scientific research far less worthy of the government's support than applied research, owing to the former's lack of perceived benefits. Other politicians castigate governmental support of applied research on the ground that it would better be left to the private sector.

Amidst the choir of dissonant voices, Congress has been concerned to determine what is being called "a new contract between science and society" for the post-Cold War era.[2] According to the late Representative George E. Brown Jr. (cited in Garfinkel and Weiss 1999: 6), "a new science policy should articulate the public's interest in supporting science—the goals and values the public should expect of the scientific enterprise." The way science is supported, the motivation for such support, and the priorities of selecting research fields and individual projects for funding are all likely to change, with consequences that may well test the high standing that American science has achieved since the end of World War II.[3]

In the prevailing policy climate of the United States, which favors market mechanisms over government intervention, basic science actually has the advantage that it can be presented as a textbook example of "market failure." Because curiosity-driven basic research is typically long-term and high-risk, and in some cases also expensive, and because its benefits typically cannot be appropriated by the organization that has carried it out, private firms (with a few notable exceptions) hesitate to make significant investments in it. In addition, national defense, which increasingly depends on science and advanced technology, has essentially only one "customer": the government.[4] This line of reasoning suggests that, to thrive, basic research needs the active support of the government. But how much basic research do we need, and on what? In this situation of widespread soul searching, our aim is to propose an imperative for an invigorated science policy that adds to the well-established arguments for government-sponsored basic scientific research. In a novel way, this imperative couples basic research tightly with the national interest.

First we shall characterize the two familiar main types of scientific research that have been vying for support: basic ("curiosity-driven") research and applied ("mission-oriented") research. These two notions are indeed useful; however, this dichotomy, which has dominated thinking about research policy, harbors two crucial flaws. The first of these is that in practice the two contenders are usually tightly interacting and collaborating, rather than being clear-cut antitheses and natural rivals in claims for support, despite what the most fervent advocates of either type may think. The history of science unmistakably teaches that most of the great discoveries that ultimately turned out to have beneficial effects for society were motivated by pure curiosity, without thought for such benefits. Equally, the history of technology recounts magnificent achievements in basic science by researchers who embarked on their work with practical or developmental interests.[5] The second important flaw in the usual antithesis is that these two widespread and ancient modes of thinking about science, pure and applied, have tended to displace and derogate a third mode—Jeffersonian Science—that deserves the attention of researchers and policy makers.

We do not advocate replacing the other two modes with Jeffersonian Science. Science policy should never withdraw from either basic or applied science. It would be particularly pernicious if federal funds were reassigned from basic science to applied science under the supposition of a quicker "social return" from the latter. Paul Berg and Maxine Singer (1998) remind us that many of the fundamental breakthroughs in science have occurred on the scientific fringes in unpredictable ways. These advances have often led to the most practical improvements, and this underscores the importance of supporting research whose relevance is not immediately obvious. The National Science Foundation has also emphasized this point from time to time—see, e.g., NSF 1980a.

While we agree that there must remain a prominent place in federal science policy for basic research with unknown societal benefits, as well as for applied research, we argue that an integrated framework of science policy explicitly including Jeffersonian Science would contribute greatly to mobilizing support for science and to propelling both scientific and societal progress. Jeffersonian Science is not an empty dream. Several federal initiatives and programs can be characterized as Jeffersonian in effect, although they did not use an explicit and overarching Jeffersonian rationale. In particular, Jeffersonian Science has had a fascinating tryout at the highest level, now almost forgotten but eminently worth remembering and reviving. We shall document this precedent after a brief exposition of our conceptual framework.

The Newtonian Mode of Scientific Research

The concept of pursuing scientific knowledge for its own sake, letting oneself be guided chiefly by the sometimes overpowering inner need to follow one's curiosity, has been associated with the names of many of the greatest scientists, but most often with that of Isaac Newton. His *Principia* (1687) may well be said to have given the seventeenth-century scientific revolution its strongest forward thrust. It can be seen as the work of a scientist motivated by the abstract goal to achieve, eventually, complete intellectual "mastery of the world of sensations" (Max Planck's phrase). Newton's program has been identified with the search for *omniscience* concerning the world accessible to experience and experiment, and hence with the primary aim of developing a scientific world picture within which all parts of science cohere. In other words, it is motivated by a quest for better scientific knowledge.

That approach—which embodies most closely the stereotypical notion of "ivory tower" science—can be called the Newtonian mode. Within the much more limited horizon of ambition of the specific projects undertaken by most research scientists, the term also characterizes the life of laboratories and theoretical centers in the largest part of academe. Today's academic researchers generally do not have as an immediate goal a grand unification, such as Newton hoped for, yet they are still committed to Newton's quest for omniscience while being satisfied to prepare one more element for the building of what sometimes has been referred to as the Temple of Science. The ascent to the highest Acropolis, from which one will be able to look over all the sciences, is left to relatively few; however, there is intense activity in the plains and foothills below, without which the upward movement would never be possible.

In all this past and present activity in the Newtonian mode, the hope for practical and benign applications of the knowledge gained in this way is a real but secondary consideration. Such applications have been and are occurring constantly, and they help explain why citizens of the United States, on the whole, think of basic science as leading far more likely than not to beneficent uses. Initially unforeseen but eventually delivered social gains stemming from "pure" science—unpredictable spinoffs—have undergirded the implied "contract" between science and society so far.

The Baconian Mode of Scientific Research

Turning to the second of the main styles of scientific research, which is popularly identified as "mission-oriented," "applied," or "problem-solving,"[6] we find ourselves among those who might be said to follow the call of Francis Bacon. Bacon is commonly remembered for having urged the use of science not only for "knowledge of causes and secret motion of things" but also in the service of *omnipotence*—"the enlarging of the bounds of human empire, to the effecting of all things possible."[7] Even stripped of its overarching and overextended rhetoric, the latter aspect of this approach is characteristic not of those whose search proceeds without prime regard for applications but of those to whom the "effecting of all things possible" is the main prize.

More research in the Baconian mode has been carried out in industry than in academe. Unlike basic research, applied research, by definition, hopes for practical and preferably rapid benefits, and it proceeds to produce those benefits by using what is already known.

Vannevar Bush's report *Science, the Endless Frontier* (dated July 1945), which along with the Steelman Report of 1947 has been a blueprint for much of America's postwar science system, shaped people's ways of thinking about the two modes of science mentioned above. George Wise (1985: 229) notes that, starting in 1945, "an implicit argument took place in which the policymakers based their policies on a simple but incorrect model [of the relationship of science and technology], while the historians began to gather the pieces for a new model not yet built." Wise continues:

"The oversimplified model favored by the policy makers depicts science and technology as an assembly line. The beginning of the line is an idea in the head of the scientist. At subsequent work stations along that assembly line, operations labeled applied research, invention, development, engineering, and marketing transform that idea into an innovation. A society seeking innovations should, in the assembly-line view, put money into pure science at the front end of the process. In due time, innovations will come out of the other end."

The assembly-line account of innovation, commonly attributed to Vannevar Bush, is of course only a simplified caricature of his real views.[8] Bush was above all a superb engineer, and thus his approach was more complex than simply linear. This becomes particularly evident when one reads a lesser-known "Bush Report." In 1949, James Bryant Conant, president of Harvard University, had asked Bush to chair a committee that was to make recommendations on how to spend a sizable bequest for engineering. The resulting report, released in 1950, advocated "extensive and high calibre research devoted primarily to bridging the gap between science and practical affairs, by seeking new science which may be applied, and more effective means of applying science in a practical and economic manner to the needs of mankind" (Bush 1950: 9). On the other hand, Bush clearly feared that too much pressure for applied research "invariably drives out" pure research (1945: 83) and that not enough basic science would be performed in the United States to provide the "scientific capital [that] creates the fund from which the practical applications of knowledge must be drawn." He characterized basic research as "the pacemaker of technological progress" (ibid.: 19).[9]

By 1994, when the Clinton administration released a report titled Science in the National Interest, the notion of an assembly-line relationship between basic and applied research, seen as competing for funds, had been superseded: "We depart here from the Vannevar Bush canon which suggests a competition between basic and applied research. Instead, we acknowledge the intimate relationships among and interdependence of basic research, applied research and technology, and appreciate that progress in any one depends on advances in the others. . . ." (Clinton and Gore 1994: 17–18) The new metaphor for the relation of basic science to development, in the words of the report, was "an eco-system" rather than "a production line." Or, as many science policy makers now put it, we are really talking about a "seamless web." Far from being separate and distinct, the seemingly initially unrelated pursuits of basic knowledge, technology, and instrument-oriented development are, in the eye of the historian, revealed in today's practice to be a single tightly woven fabric.

In an AAAS Plenary Lecture given on February 13, 1998, Harold Varmus, director of the National Institutes of Health from 1993 through 1999, eloquently acknowledged the close association of biomedical advances with progress in the more "basic" sciences: "Most of the revolutionary changes that have occurred in biology and medicine are rooted in new methods. Those, in turn, are usually rooted in fundamental discoveries in many different fields. Some of these are so obvious that we lose sight of them—like the role of nuclear physics in producing radioisotopes essential for most of modern medicine." Varmus went on to cite a host of other examples that outline the seamless web between medicine and a wide range of "basic" science disciplines.

In light of the insights of Varmus and others, what can a scientist or a science administrator do to encourage the scientific enterprise? One answer is to ensure the flourishing of the most promising projects in both the Newtonian and the Baconian mode—not least because, as history shows, they will soon interpenetrate and "trade" their respective results, for the betterment of projects of both kinds. This may not be equally true in every area or specialty; for example, it is less so in "pure" mathematics, and it is much more so in biomedical research (where the resources and results of pure and applied research are often connected in a fruitful way from the beginning). Indisputably, the trained intuition of scientists and policy makers will still have to be used to make sound decisions in each specific case.

The Jeffersonian Mode of Scientific Research

There is a third mode of research that does not quite fall under the headings we have discussed so far, and it may open a new window of opportunity, not least in Congress and the federal agencies. Whereas discipline-oriented, purely Newtonian research seeks knowledge essentially regardless of future applications, and Baconian research looks essentially for practical applications of science already known, Jeffersonian research is a conscious combination of aspects of the two previous modes. It is best characterized as follows: The specific research project is motivated by placing it in an area of basic scientific ignorance that seems to lie at the heart of a social problem.[10]

Its main goal is to remove basic ignorance in an uncharted area of science and thereby to attain knowledge that will have a fair probability—even if it is years distant—of being brought to bear on a persistent, debilitating national (or international) problem.

An early and impressive example of Jeffersonian research (and a reason for our choice of that term) was President Thomas Jefferson's decision to launch the Lewis and Clark expedition. Jefferson, who declared himself most happy when engaged in some scientific pursuit, correctly understood that the expedition would have *two* results. It would serve basic science by bringing back maps, samples of the unknown fauna and flora, and observations on the native inhabitants of western North America. At the same time, however, Jefferson realized, such knowledge would eventually serve such practical purposes as establishing relations with the indigenous peoples, and that it would further the westward expansion of a burgeoning population. Shrewdly, Jefferson persuaded Congress to back the expedition by emphasizing its commercial potential. However, to the Spanish authorities in charge of some of the territories through which Lewis and Clark would have to pass he stressed that it was a scientific mission. And in fact that mixture of motivations was correct. The expedition implied a dual-purpose style of research: basic scientific study of the best sort (suitable for a Ph.D. thesis, in modern terms) having no sure short-time "payoff" but targeted in an area where there was a recognized problem affecting society.[11]

This third mode of selecting the site for scientific research creates the opportunity to make public support of all types of research more palatable to policy makers and to taxpayers. It is, after all, not hard to imagine basic-research projects that may reasonably be expected to lead, in due course, to the alleviation of well-known societal dysfunctions. Even the "purest" scientist is likely to agree that much remains to be done in cognitive psychology, in the biophysics and biochemistry involved in the process of conception, in the neurophysiology of the senses, in molecular transport across membranes, and in the physics of nanodimensional structures. In time, it is plausible to expect, the results of such basic work will give us a better grasp of complex social tasks such as, respectively, childhood education, family planning, improving the quality of life for handicapped persons, designing food plants that can use brackish water, and improving communication devices.

Other research sites for the Jeffersonian mode might include the physical chemistry of the stratosphere; the complex and interdisciplinary study of global changes in climate and in biological diversity; the part of the theory of solid-state physics that makes improving the efficiency of photovoltaic cells still a puzzle; bacterial nitrogen fixation and the search for symbionts that might work with plants other than legumes; the mathematics of risk calculation for complex structures; the physiological processes governing the aging cell; the sociology underlying the anxiety of some parts of the population about mathematics, technology, and science itself; and the anthropology of ancient tribal behavior that still appears to be at the base of genocide, racism, and war.

Jeffersonian arguments are already being made from time to time and from case to case; that is, problems of practical importance are occasionally used to justify federal support of basic science. The most advanced government unit in this respect is the National Institutes of Health, which have been very successful in garnering federal support for basic research by linking it to the prime societal concern of combating disease. In addition, current basic research in atmospheric chemistry and climate modeling sponsored by the National Science Foundation is linked to the issue of global warming, and support for plasma science by the US Department of Energy is justified as providing the basis for controlled fusion. Yet the federal research effort seems to lack an overarching theoretical rationale and an institutional legitimization of Jeffersonian Science. In constructing such a generalized Jeffersonian agenda, the NIH may well serve as an exemplar for winning strategies that might also work in other areas. An explicit Jeffersonian agenda can also be easily understood by the public. It defuses the increasing charges that science is not sufficiently concerned with "useful" applications, for that third mode of research is located precisely in areas that are relevant to national (and international) welfare.

A variant of Jeffersonian Science that exists mainly in industry should be mentioned. Not all technological research is the same; it ranges from the highly structured application of known knowledge to what could be called basic technological research (Branscomb 1998). Much of the latter, which in our framework corresponds to Jeffersonian Science, has been conducted in the big research laboratories of major companies, such as IBM and (in earlier years) the Bell Telephone Company. It can be very basic research in terms of the creativity of the process and the uncertainty of the outcome; yet it is expected to benefit the company, and it takes place in broad research areas in which the company perceives an interest. A major motivation for industrial companies to support basic technological research is to better understand the likely evolutionary track of technology and thus to improve the quality of their technological decisions.

A spectacularly successful example of basic technological research is a project, undertaken in the late 1940s by John Bardeen, Walter Brattain, and William Shockley, that emerged from the wide-ranging science program at the Bell Laboratories and led to the invention of the transistor. The work of those three scientists, while advanced basic scientific knowledge, was primarily "problem focused," and (in contrast with the much more open-ended schedule for researchers in the Newtonian mode) a reasonably early practical payoff was expected. The case of the transistor and its transformation into the microchip also illustrates that the intended and unintended changes brought about by progress in engineering and technology are apt to penetrate fairly quickly into fields as varied as medicine, education, and financial transactions.

The current interest in rethinking science and technology policy beyond the confining dichotomy between basic and applied research has spawned some efforts related to ours. The linkage of basic research and the societal interest appeared in what Donald Stokes (1997) called "Pasteur's Quadrant," which overlaps to a degree with what we have called the Jeffersonian mode. Our approach also heeds Lewis Branscomb's (1995, 1998) warning that the degree of importance that utility considerations have in motivating research does not automatically determine the nature and fundamentality of the research that gets carried out. Branscomb appropriately distinguishes two somewhat independent dimensions: *how* (i.e., the character of the research process itself, ranging from basic to problem-solving or applied) and *why* (the motivation of the sponsor, ranging from seeking knowledge to reaping concrete benefits).[12] For instance, a basic-research process, which for Branscomb (1998: 115) comprises "intensely intellectual and creative activities with uncertain outcomes and risks, performed in laboratories where the researchers have a lot of freedom to explore and learn," may characterize research projects with no specific expectations of any practical applications, as well as projects that are clearly intended toward application. Branscomb's category of research that is both motivated by practical needs and conducted as basic research is very similar to our Jeffersonian Science.[13]

We can now summarize our theoretical framework in a triangle of Newtonian, Baconian, and Jeffersonian science.

Jeffersonian
why: societal need
how: basic research

Newtonian
why: need for scientific knowledge
how: basic research

Baconian
why: societal need
how: applied research

For many scientists, a Jeffersonian agenda would be liberating. The scientists who intend to do basic research in the defined areas of societal interest are shielded from pressures to demonstrate the social usefulness of their specific projects (pressures that otherwise are probably growing) in their grant applications. Once those areas of interest are determined, the awarding of research grants can proceed according to standards of merit that are internal to science.

Strong public support for science induced by a visible and explicit Jeffersonian agenda is likely to generalize and to transfer to other sectors of federal science policy. Even abstract-minded high-energy physicists have learned the hard way that their funding depends on a generally favorable attitude toward science as a whole. Moreover, they too can be proud of the use of the campus cyclotron for cancer treatment, of the production of radioisotopes, of nuclear magnetic resonance, and of synchrotron imaging. Nor should we forget the valued participation of pure theorists in the President's Science Advisory Committee, in JASON (a committee of distinguished scientists who advise the government), and in other important government panels, or their sudden usefulness—with historic consequences—during the world wars. From every perspective, from the purely cultural role of science to national preparedness, even the "purest" scientists can continue properly to claim their share of the support given to basic science. But the size of the total can more easily be enlarged by the change we advocate in the public perception of what basic research can do for humankind.

The Press-Carter Initiative

In the late 1970s, a fascinating and now much neglected attempt was made at the highest level to attend to what we have described as the third mode. We discuss it here as one possible example (among many) of how the Jeffersonian approach may be implemented. President Jimmy Carter set this attempt in motion in 1977 when, as Frank Press later wrote (1978a: 740), he "queried the Cabinet members on what they thought some of the important research questions of national interest were." The basic rationale behind the project was what Press (personal communication) called the "dual role" approach to improving both government and science. On the one hand, the intended role of the initiative was to make the federal government carry out its overall mission more effectively. On the other hand—and government officials could view this as a welcome side effect—the initiative strove to ensure that the United States was and would remain the world's leader in science. Thus, Press's "dual role" approach has close similarities to what we call Jeffersonian Science.[14]

The story of the Press-Carter Initiative is eminently worth remembering when the "contract between science and society" is being reevaluated. What happened during the 1970s provides an illuminating precedent (and, one may add, a glimpse into the intricate, complex, and lengthy process of arriving at results within the federal government). Owing to policies begun in the last part of the Johnson presidency and continued during the Nixon presidency, federal support for research and development, and for basic research in particular, had decreased substantially from the post-Sputnik peak, and the science infrastructure had run down to an ominous extent (Brooks 1973: 110).[15] The Ford administration began to reverse this trend, and the Carter administration undertook sustained efforts to boost federal support for basic science. Carter proclaimed in his 1979 State of the Union message that "scientific research and development is an investment in the nation's future, essential for all fields, from health, agriculture, and environment to energy, space, and defense" (cited in Barfield 1982: 12). Carter relied on Frank Press to produce the rationale for strengthening the federal science programs.

At the start of the Carter administration, Press, a distinguished geophysicist, was selected as the president's science adviser and as director of the newly established Office of Science and Technology Policy (OSTP).[16] He became a major force in making the revival of federal support for science a high priority of that administration. As a part of this policy drive (summarized in Press 1978a, Press 1981a, Press 1981b, and Omenn 1984, and fully documented in Press's papers, archived at the Massachusetts Institute of Technology),[17] Press surveyed government agencies in an attempt to obtain a list of basic-research needs that, if successfully met, might help the various agencies to fulfill their essentially practical missions on behalf of the American public. Press's survey thus exemplifies the preparation of a Jeffersonian agenda.

Press's survey could not have succeeded without the groundwork that previous presidential science advisers, especially Jerome Wiesner, had laid by injecting a great amount of scientific expertise into the federal government (Press, personal communication). Only with many highly qualified scientist-administrators working in the various government departments and agencies could the issues be understood and the relevant research questions formulated.

Our account of this initiative is based primarily on the documents archived at MIT. The archival research was augmented by correspondence and interviews[18] with several persons who had been involved in the events or who were knowledgeable on the topic: William Blanpied, Lewis Branscomb, Harvey Brooks, Daryl Chubin, Donald Fredrickson, Robert Frosch, Gilbert Omenn, Frank Press, Andrew Sessler, and Philip Smith.

Precursors

Although the Press-Carter Initiative was an innovative effort to couple basic research and the national interest, the general debate about the role of the federal government with respect to basic research had a long history, and there had been various attempts at grappling with the issue. Thus, before chronicling the Press-Carter Initiative, let us briefly review what could be considered its precursors.

In many respects, the period from the late 1950s to the late 1960s—the post-Sputnik era—was the "golden age" of American basic research, an age shaped largely by Vannevar Bush's blueprint and fueled by an unprecedented and ever-expanding stream of federal funds flowing from Cold War concerns. But even in that heyday of the support of basic science there was periodic grumbling about the wisdom of spending federal money on basic research (Omenn 1985). In 1960, the American Association for the Advancement of Science's Committee on Science in the Promotion of Human Welfare described an ongoing debate about governmental support for basic research: "The basic difficulties seem to be the absence of any overall rationale in the support of science and the overemphasis on projects that give promise of immediate practical results." ("Science and human welfare," *Science* 132, July 8, 1960: 68–73) The President's Science Advisory Committee (under President Dwight D. Eisenhower) addressed the issue of basic research in a statement titled Scientific Progress, the Universities, and the Federal Government, released in November 1960.

One of the assignments given to the National Science Foundation by its founding act (1950) was evaluating the science programs of other government agencies with a view to formulating a coordinated "science program" for the federal government as a whole. A nascent and still tiny agency such as the NSF was very reluctant to undertake this, and in 1962 President John F. Kennedy gave the task to his science adviser, Jerome Wiesner. Wiesner, in turn, asked Hugh Loweth of the Bureau of the Budget to bring together a government-wide task force consisting of the principal science officers of each of the federal agencies supporting science. Wiesner also asked Harvey Brooks to co-chair this task force as an outside adviser (as a member of the President's Science Advisory Committee).

It turned out that these science officers had never gotten together in such a way before, having instead negotiated individually with the Bureau of the Budget on behalf of their respective agencies, but never in the light of a government-wide science plan. This was a time when there were mutterings in Congress (e.g., by Senator Hubert H. Humphrey, D-Minnesota) about founding a Department of Science. The discussion within the task force was useful, but it fell far short of coming up with a unified, coherent plan, or even agreeing that such a plan was necessary or desirable. "By and large," Harvey Brooks remembers (personal communication), "our exercise was much less productive than the Press-Carter exercise apparently was. We found that the question was just too novel at that time to elicit very meaningful answers." The "science budget" was dominated by big-ticket items such as the burgeoning Apollo program and ballistic missile programs, and projected R&D budgets tended to fall off a cliff with the projected moon landing at the end of the decade. At the same time, the program of the NIH was expanding so rapidly that it dwarfed all the other programs of basic and applied research. Perhaps the most useful function of the exercise was to make some of the health professionals aware of new developments in the physical sciences that might have applications in biology and medicine. It did not (as some may have hoped) lay the groundwork for a Department of Science, nor did it even suggest that such a department might make sense in the context of US science policy.

Soon after this unproductive initiative, the Committee on Science and Astronautics of the House of Representatives undertook an attempt to help devise a sound rationale for federal support of science. In December 1963 it contracted the National Academy of Sciences—in the "first contract ever entered into by Congress and the Academy" (NAS 1965: v)—to conduct a thorough study of the matter. The NAS had already begun a project on its own. At the request of its membership at the annual meeting in April 1963, it had initiated a study (funded by the Ford Foundation) that in 1964 would release a report titled Federal Support of Basic Research in Institutions of Higher Learning.[19] For the House committee, the NAS would produce two substantial reports on the topic (Basic Research and National Goals in 1965 and Applied Science and Technological Progress in 1967),[20] but these efforts would not forestall the retrenchment of research funding in the late 1960s.

President Lyndon B. Johnson conspicuously emphasized the importance of tangible (and political) benefits when he declared during a visit to the National Institutes of Health in 1966 that "presidents . . . need to show more interest in what the specific results of research are—in their lifetime and in their administration" (cited by Omenn 1985: 1110).[21]

In 1966, when the Office of Naval Research (a major funding agency) sponsored a convocation on Research in the Service of National Purpose to celebrate its twentieth anniversary, the signs that the golden age was waning could not be overlooked. Brooks (1966: 33) warned the convocationalists: "We find ourselves today in a period of stock-taking, a time of pause and a time of soul searching. Perhaps, for the first time since the war, the assumptions on which our science policy for the past twenty years have been based are being seriously questioned and even challenged."

A few months later, the US Department of Defense released a report from Project Hindsight (Sherwin and Isenson 1967)[22] examining scientific and technological innovations that advanced the development of a number of sophisticated weapon systems. That report concluded that only 0.3 percent of these innovations had come from "undirected science." Because this study was likely to fuel criticism of the scientists' favorite doctrine of "unpredictable spinoffs," it created quite a stir in the scientific community.[23]

Beginnings

The process by which the final result of the Press-Carter Initiative was achieved was complex, with fascinating and surprising turns and timely lessons. The initiative also did not happen out of the blue. After the nadir of the governmental science policy system was reached in 1973, trends in the opposite direction soon appeared, and to some extent those trends paved the way for the Press-Carter Initiative.

The state of science and technology had drawn some concern during the Ford presidency. In 1975 Vice-President Nelson Rockefeller and NSF Director H. Guyford Stever (who, after the abolition of the presidential science advisory structure, also assumed the role of a curtailed science adviser without direct access to the president) initiated two science advisory committees. According to Philip Smith (personal communication), one purpose of setting up these committees was to assuage a science community that was still deeply shocked by Nixon's dismantling of the President's Science Advisory Committee and the Office of Science and Technology. A second purpose was to build an inventory of prospective agenda issues for a science office, once one could be reestablished. One of the committees (chaired by William O. Baker) was devoted to basic research; the other (chaired by Simon Ramo) was devoted to applied research (Baker 1993: 167; Seitz 1994: 337).

The 1976 legislation that created the new Office of Science and Technology Policy also established the President's Committee on Science and Technology and directed it to conduct a comprehensive survey of the federal science and technology effort within 2 years. President Ford appointed Simon Ramo chairman and William Baker vice-chairman of PCST. In January 1977, in his last budget message to Congress, Ford cited basic research, along with defense, as requiring a real budget increase in the coming year (Barfield 1982: 1).

Although President Carter dissolved PCST shortly after taking office, that committee left a legacy in the Carter years. Heeding the report that PCST drew up before disbanding, the Carter administration turned to examining the adequacy of the basic-research programs of federal mission agencies (memo, Frank Press to William Raney, March 1, 1977). A Basic Research in Mission Agencies Steering Group was set up. Two commissions were created, one to focus on the Department of Energy and one on the Department of Defense. The research programs of the Department of Agriculture were also studied. In reaction to the expected passage of the 1977 Agriculture Act, which designated the Department of Agriculture as the lead agency for human nutrition research, an interagency task force (including about twenty agencies) was established in June 1977. Chaired by Gilbert Omenn of OSTP, it produced a Research Agenda for Human Nutrition Research that emphasized the criterion of "researchability" of proposed basic-research and applied-research projects (Omenn, personal communication). Toward the end of Carter's presidency, a commission to examine basic research within the Department of Transportation was added.

Against this background, we can now focus more specifically on how the Press-Carter Initiative evolved. According to Thaddeus Trenn (1983: 117), whose comments were based on his interviews with H. Guyford Stever, the Carter administration, in its early days, was somewhat ambiguous or even "quite 'negative' toward scientists." Trenn (ibid.: 118) further noted that at one point, "according to [William O.] Baker, Carter had decided to abolish the entire science advisory function." Although the president (holder of a B.S. degree from the US Naval Academy) considered himself a man of science, he directed some animosity toward the academic research establishment.[24] Philip Smith (personal communication) remembers "early negative meetings with the president in which he and his staff from Georgia exhibited hostility to universities and university research." Particularly "disastrous" was a meeting Carter had with university presidents, "in which they were somewhat whiny, and Carter lectured them on the basis of his experience as Georgia governor."

While Carter was ambivalent at first, Vice-President Walter Mondale played an important role in getting things started. In early 1977, Mondale too met with a group of university presidents who desired a small shift in the federal R&D budget toward basic research. Mondale, more agreeable than Carter to the university presidents' concerns, suggested that the OSTP be put in charge of such a realignment of funds (letter from Press to Charles Slichter, March 31, 1977). He also requested from Frank Press a memorandum on basic research, which Press sent off on June 10, 1977. In that memorandum, Press summarized the case for federal funding of basic research. He began by emphasizing that federal support for basic research had an extensive and successful tradition. Among the success stories he cited agricultural research after passage of the Land Grant College Act, aeronautics research, and basic research that had furthered the war effort in World War II in multiple ways. "The findings from basic research," Press added, "while seemingly long in reaching society by way of new products and services, have been shown to be the most important *overall* source of new ideas." Press noted that approximately 30 percent of the growth in the gross national product between 1950 and 1970 was attributable to spinoff effects from basic research.[25] Yet "federal support for basic research declined by about 20% between 1967 and 1975 in terms of constant dollars. This is due to the shift of mission agencies away from basic research (Mansfield amendment[26]) and the inroads of inflation. . . . If the 1967–75 trends had continued, the world's finest research machine, the American research university, would gradually wind down."

There was an additional factor—not mentioned in the memorandum—in the decline of federal support for basic research: the growing volume of activity on the "second ivory bridge" of scientists' public-interest organizations, where vocal protesters railed against what they saw as a deformation of the research universities in the interest of questionable military and political pursuits, such as the Cold War and the Vietnam War (Kevles 1987; Leslie 1993; Nelkin 1972).

Addressing the president's "concern over conflict of interest and 'old boys' club' grant decisions," Press's memorandum went on to point out that the peer-review-based allocation mechanism for science funding was basically sound, but that high-ranking federal science officials were looking into ways of further improving it.

After noting that in the United States 70 percent of basic research was funded by the federal government, Press put the size of these funds into perspective. Basic research constituted approximately $3 billion of the $28.5 billion for federally financed R&D included in the 1978 budget. Thus, the memorandum concluded that "overall, it need not be a budget threat. . . . A 1% transfer of development funds [to basic research] is equivalent to a 10% increase in basic research funds."

Basic Science and the Budget

On July 21, 1977, having received the aforementioned memorandum, Vice-President Mondale sent a memorandum to President Carter that included a digest of Frank Press's arguments for the federal support of basic research. Mondale recommended, among other things, that Press and Deputy Director of the Office of Management and Budget Bowman Cutter undertake an initial study of the issue that would result in a report and recommendations to the president. This study would include input from university presidents and from heads of federal R&D agencies. The memorandum, preserved in the MIT archives, bore Carter's check mark of approval; however, the president had added a handwritten comment: "Do not overemphasize university concern. I'm not interested in a 'college aid fund' concept." Clearly, Carter was inclined to support basic research but was still somewhat wary of the universities.

With a favorable signal from the top, the government machinery now shifted into gear. The planning process for the fiscal year 1979 budget was the principal vehicle for transforming Carter's concerns for basic research into concrete policy. This strategy depended crucially on the cooperation of the Office of Management and Budget. OMB Director Bert Lance wrote the following to the heads of executive departments and agencies in a letter dated August 15, 1977:

> The President has expressed his interest in having Federal departments and agencies examine their research and development programs to assure an appropriate balance between basic or long-term research and shorter-term applied research and development. The President is particularly concerned with the identification of critical problems currently or potentially faced by the Federal Government where basic or long-term research could assist in carrying out Federal responsibilities more effectively or where such research would provide a better basis for decision-making. . . .
>
> We are asking that in the context of developing your 1979 budget you identify whether there are specific problems in your area of responsibility that might be better addressed through basic research and then use the results of your review to determine whether available resources can be better applied to basic or long-term research associated with those problems. . . .
>
> There is a tendency to defer needed basic or long-term research to meet more pressing near-term problems. We urge that in developing budget proposals for your agency you take a balanced view in dealing with your R&D programs and be sensitive to this tendency. . . .
>
> The results of your review should be reported along with the information about R&D and basic research programs which is submitted as a routine part of your agency's budget preparation. . . . The data in that section should be supplemented with a narrative description of the problem areas identified as outlined above, and should identify any changes proposed in your agency's R&D programs to focus agency resources on long-term research related to these problem areas. . . . A report on your review should be submitted to OMB not later than September 30. . . .

Ted Greenwood (1986: 33), who worked in the OSTP under Frank Press, commended Press's strategy of forming an alliance with the OMB early on through private discussions with top OMB officials. Thus, the basic-research initiative could be presented to the government departments with the clout of the OMB, which far exceeded the clout of OSTP or that of the science adviser. Philip Smith (personal communication) also told us that "our good working relationships with OMB, especially Bowman Cutter and Hugh Loweth, enabled us to effectively 'work the system' with cabinet officers etc." Additional evidence of the alliance between the OSTP and the OMB was the recruitment of one of Press's deputies, the physician and geneticist Gilbert Omenn, to serve as an associate director of the OMB for the final year of the Carter presidency.

The request for "narrative description[s] of the problem areas identified" provided an early inkling of the comprehensive catalog of research questions that was to be compiled at the OSTP in early 1978.

In a 1978 article in *Science* titled "Science and Technology: The Road Ahead," Press (1978a: 740) summarized the complex process:

> [It] started during a recent preview when certain issues were identified in the Office of Management and Budget (OMB) planning sessions with the President. Subsequently, there were a number of meetings in which OMB and OSTP [Office of Science and Technology Policy] met with leaders in science and engineering from universities, industry, and the government to review their impressions of trends, issues, and alternatives. We also worked with the Vice-President, Cabinet members, and the heads of NASA and NSF.

In the series of small seminars with the various interested constituencies, a group of industrial R&D vice-presidents had what may have been the most crucial impact. According to Omenn (1984: 13), "the industrialists were strong supporters of university-based basic research, and their advocacy was especially helpful, since they were viewed as not benefitting directly." The fact that the industrial science community backed academic basic research was considered likely to allay the president's suspicions about the universities.

The Cabinet

On November 11, 1977, Frank Press and Bowman Cutter sent their report to the president. Endorsed by Secretary of Defense Harold Brown, by Secretary of Health, Education and Welfare Joseph Califano, and by Secretary of Energy James Schlesinger, and by Stuart Eizenstat, the president's adviser on domestic policy, the report recommended real growth in the budget for basic research and specified the targets for that growth:

- opportunities for bringing young scientists into the system
- renewal of equipment (competitive proposals, not formula grants)
- encourage universities to explore organizational changes to improve their ability to conduct research
- encourage universities to contribute more effectively in fields of *high priority interest* to the government such as energy, environment, natural resources, technological assistance for developing countries. Examples of appropriate research questions are given in Tab A.
- encourage projects important for national productivity or national security reasons
- reduce paperwork and take steps to reduce administrative costs and improve research efficiency.[27]

To soothe Carter's known dislike of anything that looked like a "college aid fund," Press and Cutter explicitly stated: "This will not be a college aid program, but a means for the Government to improve its ability to obtain maximum return on its research investment."

On November 14, 1977, the initiative reached its decision point in a cabinet meeting at which research policy was discussed on the basis of the Press-Cutter report. The report was supported by an appendix ("TAB A"—see appendix B below) that listed 17 "examples of important research questions of national interest." This list provoked Carter's curiosity. On his copy of the Press-Cutter report, Carter underlined the passage referring to TAB A and added a marginal comment. The minutes of the cabinet meeting record that "the President commended a recent memorandum by Science Advisor Frank Press on the need to focus research and development efforts on national needs."

A few days later, Carter chose the occasion of presenting the National Medals of Science to announce publicly his commitment to increasing federal support for basic research (Omenn 1984: 13). This became a major theme of the Carter presidency. In a 1979 address to the National Academy of Sciences, for instance, Carter reiterated that in the past "research that seemed to promise a quick payoff was more amply funded, while support of basic research was allowed to decline" and "the future of our scientific and technological primacy was put at risk." "I came to office determined to reverse that dangerous, shortsighted trend," he continued. "And today I reaffirm to you my commitment to basic research, the bedrock of our scientific and technological future." (quoted at ibid.: 10)

Carter asked Press to solicit additional examples of basic-research questions whose pursuit might be important to the needs of the country. Press immediately sent out a memorandum to that effect to "Vice President, Secs. of State, The Treasury, Defense, Attorney General, Secs. of The Interior; Agriculture; Commerce; Labor; Health, Education, and Welfare; Housing and Urban Development; Transportation; Energy, Admin., Environmental Protection Agency, Chairman, Council on Environmental Quality." The members of the Federal Coordinating Council for Science, Engineering, and Technology, the director of the National Institutes of Health, and the Commissioner of Food and Drugs were sent the Press-Cutter report "for your information" but without the request for additional questions; they were thereby alerted to prepare for inquiries within their departments from cabinet-level officials. In a personal communication, Gil Omenn recalled: "We wanted to be sure they knew what we sent their 'bosses,' the Cabinet officers, so they would get involved—which they did."

Many of the additional suggestions for basic research came, as Frank Press emphasized (personal communication), from the many first-rate scientists working in various government units. Their ideas were fed into a review process, and the top officials then presented the surviving questions to Press on behalf of their units. William Blanpied (personal communication) suspects that "the extent to which a given agency head responded with imagination and enthusiasm to Press's request may be closely related to the extent to which scientists in that agency were valued by, and connected with, upper management. . . . the failure of many departments to respond at all to Press's request suggests that their heads did not regard science as terribly important."

Government Agencies and the Basic Research Initiative

In a letter to acting OMB director James T. McIntyre dated December 9, 1977, Richard C. Atkinson, the director of the National Science Foundation (the government's flagship agency for basic research) expressed skepticism toward the other agencies' long-term commitment to basic research in universities on the ground that their more pressing concerns would tend to push basic research off their agendas. Echoing Vannevar Bush's famous dictum about applied science driving out basic science, Atkinson wrote: "Experience demonstrates quite clearly that other Federal agencies find it extremely difficult to increase their support for basic research in universities when they cannot meet their shorter range program requirements within the limited resources available to them. I am concerned that these pressures may once again severely constrain the amount of funds devoted to basic research in universities by other agencies with the result that overall Federal support for science may suffer." In hindsight, this skepticism on the part of Atkinson, whose primary impulse was probably to protect the NSF's share of the overall budget devoted to basic research, might be regarded as justified. At the time, however, the other agencies' responses to Press's initiative ranged from avid eagerness to veiled hostility.

Department of Defense

Particularly strong enthusiasm for the Press-Carter Initiative emanated from the Department of Defense. In a memorandum to Frank Press dated November 9, 1977, Deputy Under Secretary of Defense for Research and Engineering (Research and Advanced Technology) Ruth M. Davis agreed with and commented favorably on the draft of the Carter administration's policy on basic research. She emphasized that the new policy was a welcome trend reversal for the Department of Defense, which was still chafing under the lingering aftermath of the Mansfield amendment. Moreover, the department was having a hard time establishing its research agenda at the universities; Vietnam-era antagonism and resentment of the "military-industrial complex" still pervaded the climate on many campuses. "I believe," Davis wrote, "that DoD has experienced a worse setback than civilian agencies in being able to work with and exploit university research resources. Further, there is easily discernible evidence that university researchers are presently attracted to energy, environment, life sciences and other related areas at the expense of military or national security-related research. We need to counter this trend before it seriously and adversely affects our national security 'futures.'"[28]

The Department of Defense also provided seven additional suggestions in response to Press's inquiry (December 1, 1977). The first of the research questions submitted by the Department of Defense was as follows:

- Can materials be found that exhibit superconductivity at room temperature? Such a discovery would be extremely important to our energy needs as well as revolutionize all technology using electrical energy.

Thus, almost 10 years before the sensational breakthrough that occurred in this area in 1986–87 (to significantly higher temperatures than absolute zero, although not to room temperature), the Press-Carter Initiative was able to identify this field of basic inquiry as deserving support. That initiative, which ultimately was not translated into concrete policy, cannot be credited for bringing about the advance in superconductivity; nonetheless, this selection shows that the exercise was able to produce some forward-looking items.[29]

The other suggestions submitted by the Department of Defense covered a wide range, from basic-research questions whose immediate applicability was obvious to fundamental questions whose potential practical spinoffs remained unexplained. The first of the following three questions represents the former category, the other two the latter:

- Can new materials such as ceramics be developed to replace metals in high temperature situations? For example, use of ceramics in turbine blades will permit greatly increased operating temperatures, reduction of size, increase in efficiency, and reduction in fuel consumption.
- Is there intelligent extraterrestrial life? This fundamental question should be pursued as man penetrates more deeply into space.
- Are there fundamental building blocks in nature? Some recent advances have been made which indicate that even the subnuclear "particles" are not fundamental and further research is necessary to uncover the secrets of the nucleus.

Department of State

The Department of State was initially less eager than the Department of Defense. It flatly ignored the Lance memorandum of August 15, 1977. Having been questioned by Frank Press about this in early November, Patsy T. Mink, Assistant Secretary of State for Oceans and International Environmental and Scientific Affairs, issued a somewhat apologetic explanation on November 10: "Action on the Lance memorandum was not assigned to my Bureau. The two Bureaus which were involved determined that the State Department funds no basic research and hence has no budget item for it. As a result and following what is apparently common State Department practice, it was decided that the memorandum was not applicable to State and that no response to it was required. . . . I have been unable to convince the powers-that-be that a negative response should be sent."

But in the wake of the November 14 cabinet meeting the Department of State came around. Lucy Wilson Benson, Under Secretary for Security

Assistance, Science and Technology, sent Press five questions on January 4, 1978. One of these concerned disaster prediction:

- To what extent can the occurrences of natural hazards such as fire, flood, earthquake, and pestilence be foreseen sufficiently in advance to permit mitigation of their effects? The problems of prediction and of mitigation are different for each hazard, but for each, research offers promise of reducing human and physical costs.

The Department of State's questions also ranged widely, from social science to biology and geology:

- What are the factors—social, economic, political, and cultural—which govern population growth? High population growth rates in the developing countries impose an economic burden which too often exceeds the gains made by development. Social and biomedical research on safe, efficacious, and culturally acceptable contraceptives would therefore be of great benefit.
- Can microbiological research develop organisms which can convert crude organic materials, such as common cellulose, into livestock feed? The ability to convert common cellulose to feed-stock would significantly increase the availability of high-grade animal protein for human consumption.
- Can research into the processes by which mineral deposits were formed in the earth's crust be sufficiently aided by deep ocean floor investigations so that mineral resources can be more efficiently located on land or sea-bed? Research which would improve the success-rate of exploratory efforts could be of considerable advantage.

Department of Energy

On December 3, 1977, the Department of Energy contributed 16 suggestions, including one on superconductivity:

- Can materials be found which are more conductive or superconducting at higher temperatures? A room temperature superconductor would revolutionize our entire system of electrical conversion and transmission.

Other questions had to do with finding more efficient and more economical ways of producing energy and with the environmental impact of energy production and consumption:

- The economic and predictable fracture of rock is of critical importance to energy production. Obvious examples are drilling for new resources, mining of coal, oil shale and uranium, and releasing natural gas from low permeability formations. Research is needed on the mechanical behavior of rocks, in order to improve our understanding of them as engineering materials.
- How can considerations of second law [of thermodynamics] efficiencies be incorporated into energy strategies? Energy should be valued not by its amount alone,

but also by its thermodynamic quality. A significant reassessment of energy economics may be in order.

- How do catalysts work? Research on this question can lead to more economical ways to produce hydrogen and to convert coal to useful liquids and gases.
- At what rate will atmospheric carbon dioxide concentrations increase as a result of increased use of fossil fuels? What effect will increasing carbon dioxide levels have on climate? How will this change the global social, economic and political structure? How might the impact be ameliorated?

Yet the suggestions submitted by this department also reached into realms that would satisfy the "purest" research interests:

- How are the fundamental forces of nature related? Four types are currently known: nuclear (strong), electromagnetic, radioactive (weak) and gravitational. Only electromagnetism is well understood; the rest defy us to master them.
- Does an "island of stability" beyond the current periodic table or "abnormal" states of nuclear matter exist? These speculations can be tested and if found could have important consequences for nuclear energy production.

In June 1978, an OSTP working group on basic research at the Department of Energy reported that the existing program was unbalanced and recommended an increase the budget for basic energy research in the universities: "The overall basic research program of the DOE should be of a size and scope sufficient to ensure the health of disciplines of importance to the long-term development of energy technologies. Stability of funding is a key element in the development and maintenance of a research program of the highest quality. Excellence of research should be a principal concern when judgments are made on the support of specific proposals."

Department of Transportation

On January 5, 1978, the Department of Transportation submitted four suggestions. These focused fairly closely on transportation issues. For instance:

- What limitations, if any, exist on the development of full performance electric and hydrogen-powered automobiles, trucks and buses? Areas of concern here include the development of improved batteries and lightweight, reliable and maintainable hydrogen storage tanks. The basic discipline area of interest is materials. The need for continued transportation energy research is obvious.
- What are the implications of the advent of inexpensive, large capacity microprocessors on the decentralized control of large scale systems in general and transportation networks in particular? It is necessary to expand our efforts in the area of modern control theory in order to understand the feasibility of and benefits from decentralized control. A situation now exists where the hardware state-of-the-art is advancing faster than our ability to employ it optimally.

Department of Agriculture

On January 10, the Department of Agriculture submitted 33 research questions—more by far than any other department. In light of subsequent developments, one question is particularly remarkable:

• What are mechanisms within body cells which provide immunity to disease? Research on how cell-mediated immunity strengthens and relates to other known mechanisms is needed to more adequately protect humans and animals from disease.

This question, framed in 1978 as a basic-research question, was to become a life-and-death issue for millions only a few years later with the onset of the HIV epidemic. This instance illustrates that Frank Press's initiative was able *in advance* to identify a basic-research program whose potential benefits were understood *in principle* at the time, but whose dramatic significance could not have been foreseen (and might well not have been targeted in a narrowly application-oriented research program).[30]

National Aeronautics and Space Administration

NASA's response, submitted on February 14, 1978, distinguished itself from all the others by its closeness to the "Golden Age of Science" postulate of supporting basic research without linking it to concrete societal benefits (which were considered to accrue as unpredictable spinoffs). The NASA list started out in grand style with "What is the nature of the universe?" and then went on to more specialized questions: "What is the nature of life?" "What is the nature of gravity?" "What is the nature of intelligence?" "What is the nature of matter?"

Notably absent were any statements of the societal benefits that would be expected to derive from solving these questions. The reason for this, as Robert Frosch (NASA's administrator at the time) recalls, was that the NASA officials responded to the request with the Space Act of 1958—an act that explicitly mentioned among NASA's responsibilities the mission of advancing space exploration and aeronautics—in mind. In their view, NASA would have overstepped its bounds had it justified its basic research by any mission other than its own, to which the above questions were of course centrally related. According to Frosch (personal communication): "NASA didn't answer re broader applications of the basic knowledge because it was not charged with any broader societal missions. Everyone in NASA was well aware of broader application implications, but thought it wise to stick to our own territory and not assert the solution of problems with which other agencies were charged." NASA officials did in fact have several discussions of these wider issues with OMB's director, Bert Lance, who "was beginning to get the idea that NASA could be 'useful to the President and the Administration'" (ibid.). To the disappointment of NASA, however, Lance soon had to leave office, and nothing came of this NASA-OMB connection.

Aside from basic science constituting an explicit part of its core mission, NASA was also somewhat exceptional in the high degree to which it could tap into the often-forgotten fact that the pursuit of fundamental scientific advances fascinates many laypersons. NASA's activities in space exploration, in particular, were able to capture the imagination of wider segments of the American public, perhaps in part because they resonated with the frontier and pioneer mythology deeply embedded in the American psyche.

Attorney General

In a letter dated November 29, 1977, Attorney General Griffin Bell let Frank Press know that basic research was peripheral to his department's work: "Except for the research and development roles supervised by the Law Enforcement Assistance Administration, the Department of Justice has no present involvement in such basic research."

The other government departments and agencies that Press asked for additional suggestions for basic research in the national interest (Treasury; Interior; Commerce; Labor; Health, Education, and Welfare[31]; Housing and Urban Development; Environmental Protection Agency; Council on Environmental Quality) did not respond. According to Gil Omenn (personal communication), their reticence was probably attributable to their excessive caution about how the Congress might react to anything they might do that did not appear to be directly tied to their missions. HUD and EPA, which were funded by the same appropriations subcommittees that funded the NSF, were particularly circumspect. To differentiate themselves clearly from the NSF (the main agency for basic research), and to sidestep the indifference toward funding basic research that pervaded Congress, these units tended to claim that no "basic research" would be carried out under their auspices.

Philip Smith (personal communication) did not find it surprising that these units did not respond: "These departments have the dual problems of an R&D effort of varying degrees of effectiveness and cabinet leadership that generally does not have much interest in R&D." And the R was typically farther from cabinet officers' minds than the D.

National Institutes of Health

The NIH, which obviously had a large stake in research, was somewhat antagonistic to Frank Press's efforts. At the time, the NIH was involved in an attempt to redefine the principles by which the government should support health research that was somewhat analogous to the Press-Carter Initiative. That effort had been triggered by Health, Education and Welfare Secretary Joseph Califano's project to develop a five-year plan for new health-related research expenditures. Fundamental research was given high priority, and the congressional appropriations committees were eventually convinced to fund at least 5,000 investigator-initiated grants (which were considered to be on the basic-science end of the spectrum of NIH-supported research) per year.

At the start of the Press-Carter Initiative, the NIH devoted one-fourth of its funds to what was classified as basic research. The OSTP tried to increase this portion. On November 1, 1977, Gil Omenn, one of Frank Press's deputies, reported at a meeting of senior OSTP staffers that he and officials from the OMB's Health Branch had agreed to increase the fiscal year 1979 basic-research budget of the NIH from $747 million to $850 million. This amount would have to come out of other areas of the NIH budget. According to Omenn's memorandum, there was some sense of an uphill battle ahead: "The Congress is likely to earmark large funds for highly applied areas." Moreover, the NIH itself seemed to be dragging its feet. "Unfortunately," Omenn continued, "NIH has not replied to the OMB letter on basic research. [NIH Director Donald S.] Fredrickson insists that DHEW never referred the official letter to him, though I sent him a copy."

When Fredrickson did send a letter to Omenn, it signaled OSTP to back off. While explicitly stating that the NIH would remain "enthusiastic about the President's concern for 'basic' research," the letter also highlighted the problems the president's initiative would cause. The main message of the memorandum, diplomatically couched in enthusiastic language, was a veiled threat that a reshuffling of funds toward "basic" research would create a public outcry and a political backlash because it would jeopardize popular NIH programs:

The impact of reprogramming funds from "applied" to "basic" research will be felt throughout the NIH research community. Some of the effects of this reprogramming are:

• . . . a reduction in the funds available for their comprehensive research centers in areas such as epilepsy, trauma, stroke, and multiple sclerosis.

• . . . [reduce support for] programs in asthma and allergic diseases and in clinical immunology. Vaccine development and anti-viral evaluation, and clinical trials of viral-preventive or treatment substances will be delayed.

• . . . hamper the national effort to control dental caries in children.

• [At the National Institute on Aging] . . . reduce many of their "applied" efforts, particularly in epidemiology, clinical nutrition, pharmacology and social and behavioral research on life expectancy.

In the same letter, Fredrickson summarized the ongoing attempts on the part of the NIH to reconceptualize research beyond the basic/applied dichotomy. He had been skeptical about this distinction for quite a while; he thought it did not fit the realities of NIH-sponsored research in the biomedical field. In fact, Fredrickson (personal communication) considers the NIH a particularly good example of the Jeffersonian mode in action. Throughout his letter to Omenn, he put quotation marks around the words *basic* and *applied.* He also outlined the alternative classification scheme that was emerging from the NIH deliberations:

As you are aware, classification of research projects as "basic" research and "applied" research is very tenuous. Often, the classification depends on one's purpose or location on the continuum which extends from the search for better understanding of natural processes, through the validation and the application of knowledge in order to influence those natural processes, to the translation of knowledge into health care practice. We are currently attempting to redefine the areas of the continuum as science base, application and transfer to achieve a more accurate portrayal of the NIH programs. The tentative definitions of these areas are:

• Science-based—includes both the development of new knowledge about the fundamental life process ("basic" research), and the validation through targeted research of the developed knowledge, focusing on disease prevention, diagnosis and treatment ("applied" research); this research is accomplished primarily through investigator initiated research, program project and some center based activities;

• Application—application of validated knowledge to develop safe and efficacious interventions and products, generally including all clinical trials specifically indexed in the NIH Clinical Trial Index, other research of a more developmental nature, such as drug and vaccine development; and instrument development, most of which is contract supported;

• Transfer—demonstration, control, education, consensus building, and the transfer of knowledge to the health care community.

In his reply, Omenn tried to reduce the magnitude of Fredrickson's claims:

Frank Press and I reviewed your memo in detail. Only $26.8 million of the $93.0 million increase in basic research is derived from reprogramming of grants and contracts—and part of that is a shift from contracts to grants within the viral oncology program. It is not the intention of the Administration to neglect clinical research and clinical trials which give promise of significant advances, including the areas mentioned in your letter (epilepsy, dental caries in children, vaccine development, and nutrition among the aged).

But the NIH situation remained somewhat problematic for the OSTP initiative. In a memorandum to James McIntyre about the fiscal year 1980 R&D budget, dated November 16, 1978, Press wrote: "*Basic research*, leaving aside the complex situation of the NIH, is in good shape. Preliminary figures indicate +11%, though USDA and Interior fare very badly (Interior –4.2%; Agriculture +0.4%)." Omenn (1984: 14) pointed out that every year since the 1960s the presidential budget requests had shown smallish increases for the NIH because the administrations reliably expected Congress to add a substantial increment—in contrast with congressional trimming of the NSF's and other agencies' requests for research funds. Philip Smith (personal communication) noted that it was "not in the self-interest of NIH to cooperate with cross agency initiatives." Similarly, William Blanpied (personal communication) stated that the NIH had "done very well with its budgets by focusing on the importance of its work to the nation's health, rather than (like the NSF) attempting to convince Congress that basic research ultimately leads to significant practical benefits."

The budgets of the NIH were indeed remarkably stable over time, with a steady upward trend.[32] Because the NIH was already highly successful in using the "health argument" to acquire research funding (even for rather "basic" projects), it had little to gain and much to lose by joining forces with an interdepartmental drive in support of basic science.

The Master List

Overall, the OSTP staff was pleased with the results of the initiative. On January 31, 1978, an elated Gil Omenn wrote to Frank Press:

I share your excitement in reading these replies. It is remarkable that Government agencies have generated such impressive lists of research opportunities. The one-line statements of "the significance of the research" are often compelling.

> Replies have been received only from State, Defense, Energy, and Agriculture. NSF had much input in your original list.
>
> The replies are very heterogeneous. Some recite the agency research agenda (USDA and to a lesser extent Energy). Defense and State add only a few well-selected examples to your already strong list, and these examples are broader than the agency agenda. Not all are "basic."

In the same memorandum, Omenn recommended the following to Press: "Make a master list of your two pages plus about 3 more pages drawn from the others, organized by field, to append to a statement that you might make in a speech or at a press briefing—as *examples* of the kind of research-oriented thinking going on in this Administration." Here we find the inception of what was to be the major product of the Press-Carter Initiative.

Stanley D. Schneider of OSTP combined all the lists of suggestions except the one from the Department of Transportation.[33] The resulting "master list" of 87 questions was distributed in an OSTP news release to "All Science Writers" on February 13, 1978. That master list, then, contained all the basic-research questions whose solutions, in the view of the federal agencies responding, were expected to help the federal government significantly in fulfilling its mission.[34]

The master list shows that there was indeed a novel link being forged between "pure," curiosity-driven research and the US government's interest in furthering practical benefits. No longer was the link provided merely by chance and providence. At the same time, Press's model resisted public pressure to spend taxpayers' money chiefly on "useful" (i.e., applied) research and to cut back further on "useless" basic research. The new link was secured by looking for basic-research questions in areas of scientific ignorance that could be identified as obstacles to practical benefits, thus connecting basic research with probable (if not predictable) social payoffs. In effect, Press asked the government agencies to look beyond their immediate problems and their applied-research interests and to identify basic-research questions whose solutions would be likely to have a positive impact on them eventually.

This procedure, of course, selected only a subgroup of all the pure-research questions that could be conceived in a curiosity-driven mode. However, it was an innovative attempt to legitimize federal funding for basic research by linking that research to public needs. The long and varied list of questions in that astounding master list also showed that such a model would substantially preserve the scientists' own ultimate aims as well as their opportunities for obtaining federal funding for "pure research" while raising concrete hopes for the solution of problems of societal importance. Many of the selected research questions were basic enough to resonate with the intrinsic standards of good science within the scientific community, but they simultaneously corresponded to what the heads of federal government agencies (including NASA and the Departments of Agriculture, Defense, Energy, and State) considered good science at that time—good, here, in the sense of expected eventual practical payoffs. Of course, some of the questions on the list seemed more basic (e.g., "What is the nature of the universe?") than others (e.g., "How can structures be designed and constructed to be both economical and earthquake resistant?"). "The replies are very heterogeneous," Gil Omenn commented. "Not all are 'basic'."

A few of the questions (especially those submitted by NASA) did not follow the specified format, which called for an explicit link between each question and potential benefits. On the whole, however, it is striking how closely this list tied basic-research suggestions to hoped-for practical improvements of societal significance. Remarkably, a few of the basic-research questions that made the list—such as those concerning superconductivity and the human immune system—were only hypothetically relevant at the time but subsequently became vastly more urgent.

Similar Initiatives

A similar intention of linking basic research to perceived national needs underlay the 1978 reorganization of the National Science Foundation's applied-research programs, during which, under the new Directorate for Applied Science and Research Applications, a division of Integrated Basic Research was formed. Unlike RANN (Research Applied to National Needs), an earlier applied-research program that aimed to encourage and accelerate the application of *existing* basic scientific knowledge to a wide range of potential uses, the Integrated Basic Research division was formed to provide "support for basic research that has a high relevance to major problems" in selected topic areas in the basic-research directorates.

At the end of the 1960s, the NSF was under growing pressure to heed its explicit applied-research mission, which the Daddario-Kennedy Amendment of 1968 had added to the National Science Foundation Act of 1950. The

NSF's response was RANN,[35] which, including a precursor program, lasted from 1969 to 1978 (Mazuzan 1994). As an incentive, the OMB promised the NSF a budget increase of $100 million, half for the new program and half to take over projects that had to be relinquished by the Army, the Navy, and the Air Force because of the 1970 Mansfield Amendment. From its start, RANN differed from the traditional NSF programs. It was organized around designated problems that might be solved largely with known science, and it attempted to link up industry with academic research. Criticism of this new program abounded, the fiercest and most vociferous of it coming from the academic science community, which feared a subversion of the NSF that would result in a loss of support for basic research. In addition, many academic scientists resented that RANN was directed by former NASA officials who did not appear to appreciate the culture of basic science. The 1978 reorganization of the NSF's applied-research programs, which finally placated the basic-science community, could be interpreted as a move toward a more Jeffersonian approach.

A Jeffersonian approach also pervaded the Tenth Annual Report of the National Science Board of the NSF, released on August 2, 1978 and titled Basic Research in the Mission Agencies: Agency Perspectives on the Conduct and Support of Basic Research (NSB 1978). In the now-familiar vein, the report contained a collection of basic-research topics that were perceived as responding to national needs and thus meriting government support. Sixteen agencies, including the Department of Agriculture, the National Science Foundation, the Department of Housing and Urban Development, and the Veterans Administration, had contributed basic-research questions whose solutions, in their view, would benefit their missions.

Further momentum was added in 1980 when the NSF released a two-volume report titled The Five-Year Outlook: Problems, Opportunities and Constraints in Science and Technology (NSF 1980b). Again, long and detailed lists of fundamental research to be encouraged were compiled—lists that showed considerable overlap with those in the earlier report.

The activities of the National Science Foundation and the National Science Board reflected the political mood in science and technology policy that had been created in the Press-Carter period. Richard Atkinson and later John Slaughter, as NSF directors, shared the fundamental outlook of Frank Press and his OSTP staff. What distinguishes Press's list from these allied and more comprehensive efforts is the immediate link that the government agencies were requested to forge between specific basic-research questions and the national interest.

Outcome

The fate of these hopeful steps toward an agenda for basic science in the societal interest was not entirely encouraging on the first try at the highest level. To be sure, the Carter administration's emphasis on basic research expressed itself noticeably in the budget proposal for fiscal year 1979, which recommended real growth (net of inflation) in expenditures on basic research of almost 5 percent. By the time the budget was approved by Congress, however, almost nothing of the proposed growth remained, largely as a result of unexpectedly high inflation. Efforts in the following years also fell short of achieving a substantial increase in funding for basic research. A 1980 *Science* article noted that "the Administration's actual spending on basic research over the last 3 or 4 years has just stayed ahead of inflation" (Marshall 1980). More immediately urgent concerns overshadowed the Press-Carter Initiative and precluded the iterations beyond the assembly of the "master list" that would have been necessary for its actual implementation in science policy. Finally, the defeat of Jimmy Carter by Ronald Reagan in the presidential election of 1980 made the Carter administration's plans to boost spending on basic research in the early 1980s obsolete; it also squelched followups to the Press-Carter Initiative in the NSF and in other government agencies.[36]

Hampered as it was by adverse political conditions and soon terminated by the electoral defeat, the Press-Carter Initiative may seem to have been only a relatively ephemeral episode in presidential science policy. Yet it was a major effort toward institutionalizing Jeffersonian Science in the twentieth century. During the Reagan years, many of the topics touched upon by that initiative were kept alive within a different institutional setting. At the end of Carter's term, Frank Press moved on to become president of the National Academy of Sciences, and some senior members of his OSTP staff went with him. Gil Omenn, who had returned to the University of Washington, served as member and then chairman of the Committee on Science, Engineering, and Public Policy (COSEPUP) at the National Academies.[37] COSEPUP issued field reports to the new administration's science adviser and to the agency heads (at the annual request of the science adviser). Philip Smith became Executive Director of the National Research Council, a body sponsored jointly by the National Academy of Sciences, the National Academy of Engineering, and the Institute of Medicine.

The success of the Press-Carter Initiative in generating a credible list of research questions in the national interest rested on several preconditions, including close collaboration with the OMB, backing from the president, and the scientific expertise available in the agencies. After the creation of the master list, the initiative came to a halt, not because it was energetically opposed or because it was flawed, but because it simply faded away. What would have been the next steps? What would have been the mechanisms for follow-through? How would the list have been prioritized and then turned it into a concrete policy and a concrete organizational arrangement? How would the convergence among the questions submitted by different agencies have been handled? Would the NSF and the NIH have come around? How would the initiative have played out in the legislative branch?

Next Steps

A revitalization of the Press-Carter Initiative would provide a promising additional model for future science policies—one that would be especially relevant to the current state of debate and disorientation about the role of science in society and to renewed examination of the reasons for federal funding of basic research. Let us suggest two steps.

The first is to evaluate the Press-Carter Initiative and the list of questions it generated with the benefit of the hindsight gained from two decades' scientific progress. Retrospective consideration of the initiative affords a unique chance to consider the promise of a Jeffersonian style that focuses on problems of basic science whose solutions will, with reasonably high probability, benefit the societal interest. Which of the questions compiled in 1978 still appear important, and which do we now consider scientific blind alleys? Which questions have been worked on, and which have been neglected or abandoned? Which problems have been solved or have seen significant progress, and which are still intractable? Which questions reflected research that was already well underway, and which were truly forward looking? Finally, what questions should have been asked but were not? Such a comprehensive evaluation clearly exceeds the competence of any single researcher. Only a number of panels composed of experts in the various fields could provide authoritative or at least plausible answers.

The second step is to turn to the future. We urge the US government and the scientific organizations to engage in a new version of the Press-Carter Initiative that would help to enlarge the spectrum of American science. Among many other potential benefits, a federal science policy that includes a strong and explicit Jeffersonian agenda might be instrumental in attracting more women and more members of underrepresented minorities to science careers.[38] Using an appropriate institutional basis (perhaps OSTP in concert with the NSF and other agencies),[39] and benefiting from the lessons of the first trial, a revived Jeffersonian initiative would contribute decisively to a sorely needed overall framework for federal science policy.

Concrete efforts have already begun. A major one was a national conference on Basic Research in the Service of Public Objectives, held in Washington in November 2000, that explored the concept of a Jeffersonian agenda and generated positive responses from numerous leaders of the science policy community. For instance, NSF Director Rita Colwell emphasized that "the objective of connecting discovery to society is central to our work." In 2001, the Center for Science, Policy, and Outcomes, located in Washington, became the institutional home for future activities to shape a Jeffersonian agenda.

We emphasize again that we do not advocate complete replacement of the Newtonian and Baconian modes by the Jeffersonian mode. Each mode must be a part of an integrated federal science policy, and funding for all three modes must be stable and indeed growing. Yet the conscious and explicit addition of a Jeffersonian research program could powerfully address current complaints that science is not sufficiently concerned with the welfare of society. Such a program would set a clear agenda and a series of new goals for basic science in the societal interest.

3

Organizing for the Common Good: Scientists' Voluntary Public-Interest Associations

In the fifth century BC, the Roman authorities summoned Lucius Quinctius Cincinnatus, a private citizen and a farmer, straight from his plow to save an army surrounded by enemy forces. After he was appointed "dictator," this famed figure of the early Roman Republic vanquished the enemy and promptly returned to his fields, unimpressed by the lure of holding onto political power. This outstanding display of civic virtue made Cincinnatus one of the heroes of the republican spirit, and he was much venerated in the early days of American independence. In 1783 a group of officers who had fought on the American side in the Revolutionary War formed a patriotic organization called the Society of the Cincinnati, and in 1790 a city in Ohio was named in honor of that society.

Unlike Cincinnatus, the citizens who form associations for the public good do not serve as government officials, yet in an important sense they act in a Cincinnatus-like fashion. Often compelled by what they perceive as a crisis threatening the community, they enter the public realm in the spirit of civic duty, not to gain a career or personal advantage.

The penchant of private citizens for creating voluntary associations was one of the features of American public life that deeply impressed Alexis de Tocqueville in the early nineteenth century. "Americans of all ages, all conditions, and all dispositions, constantly form associations," he wrote in amazement. "Wherever, at the head of some new undertaking, you see the government in France, or a man of rank in England, in the United States you will be sure to find an association. . . . I have often admired the extreme skill with which the inhabitants of the United States succeed in proposing a common object to the exertions of a great many men, and in inducing them voluntarily to pursue them." (Tocqueville 1862: 129–130)

It is clear that the associative proclivity of Americans has persisted to present times. Our aim is to examine a specific part of this wider phenomenon: the astounding variety and multitude of public-interest associations formed by scientists in the post-World War II era. Several dozen such organizations have been founded, some famous and others more obscure but still important. A major theme of this chapter will be how scientists' post-World War II public-interest associations started out with a Cincinnatus-like posture but then changed as they became more institutionalized and professionalized.

Remarkably, the existence and the extent of the assemblage of scientists' public-service associations are not clearly perceived or greatly valued by many (perhaps most) members of the scientific community. Here we find, to quote Joseph Rotblat's 1995 Nobel Prize acceptance speech again, that "precepts such as 'science is neutral' or 'science has nothing to do with politics' still prevail." Moreover, a good number of social critics, in their writings and their academic teaching, have been painting a picture that carefully leaves out scientists' activism for the public good. Instead, their picture is dominated by the view that scientists are really interested only in getting a "blank check" from society, and that the term "moral authority of science" is a self-contradiction. (See Holton 1993 and Koertge 1998.) According to an editorial in the journal *Nature* (367, 6 January 1994, p. 6), the current trend outside laboratory walls has been resulting in "a slow—and Hollywood-assisted—erosion of [the scientist's] public image, . . . replacing it in the public mind by a money-grabbing, plagiarizing con-artist."

The derogations from prominent non-scientists, and the lack of interest among so many scientists, can be ameliorated in many ways. One would be to prepare and disseminate an account of the numerous scientists' organizations of the sort we have mentioned, including a factual report of their aims and accomplishments. In appendix D below, we have assembled profiles of a large number of scientists' voluntary associations and allied organizations. We hope that this unique compilation will afford the reader a glimpse into the relatively little-known world of citizen-scientists' organizations (thus relativizing the ivory-tower stereotype of the detached and selfish scientist), and that it will provide a useful starting point for future researchers. Though we cannot provide a more comprehensive and definitive treatment of the topic in all its details here, it is our wish to stimulate further work in this important yet neglected "research site." Research in this area will also increase understanding of the scientific community among laypersons. There already exist a few fine historical, sociological, and political science theses and other studies on specific organizations (e.g., Moore 1993, 1996; Nichols 1974; Smith 1965; Strickland 1968). More such work is needed to fill out the picture of the scientists' role in society. Not the least reason for turning to this subject now is that some of the individuals who participated in the wave of activity that took place during and after World War II are still with us.

We consider the collective efforts of the citizen-scientists a crucial second ivory bridge between science and society—a bridge that complements and rivals the governmental one. The construction of two bridges, rather than one, corresponds well to a general "balance of power" motif in American politics and society. As government increasingly harnesses science and technology for its purposes, a growing citizens' movement attempts to check the government's uses of the fruits of science and technology.[1]

We will focus on post-World War II organizations, although we fully realize that scientists' activism had already manifested itself in the 1930s and during World War II. That early wave of activism was markedly varied and internally antagonistic. For example, in the 1930s, under the influence of Marxist ideas and with J. B. S. Haldane and J. D. Bernal among the main energizers, the Association of Scientific Workers in Britain and its sister organization in the United States downgraded pure science in favor of state-planned science (Filner 1976; Kuznick 1987). In opposition to this agenda, the Society for the Freedom of Science was founded in 1941 in Britain by Michael Polanyi, J. R. Baker, and A. G. Tansley to champion the independence of scientific pursuits (Society for Freedom in Science 1953). That society also attracted many adherents in the United States, among them P. W. Bridgman.

During World War II, prominent scientists' objections to the plan to drop atomic bombs on populated areas in Japan and their warning of the need for international control of atomic energy were a source of and a model for the wave of activism among scientists that immediately followed the war. After Hiroshima and Nagasaki, old images of science such as the ivory tower and the isolated and bucolic Temple of Isis no longer seemed appropriate. Our account of scientists' public-interest associations begins at the moment of that general realization.[2]

Although our primary geographic focus will be on the United States, we will also cover some international associations in which American scientists have played a prominent part and some foreign organizations that have been in close contact with their American counterparts.

The organizations we will be discussing are distinguished mainly in that they are *voluntary* (vs. official), composed of *scientists* (vs. non-scientists), and concerned primarily with the *public interest* (vs. scientific interest and self-interest).

Voluntary Organizations

In chapter 2 we dealt with the official link between science and the public interest. The elements of this connection include the president's science adviser, the Office of Science and Technology Policy (whose precursor was called the Office of Science and Technology), and, until 1973, the President's Science Advisory Committee. On the congressional side, there was (until 1995) the Office of Technology Assessment. Moreover, scientists in responsible positions at the National Science Foundation, at the National Institutes of Health, and in other government units involved with science can and do influence policy making from within the government. Finally, there are the National Academy of Sciences, the National Academy of Engineering, and the Institute of Medicine, and there is the National Research Council (jointly sponsored by the aforementioned three institutions); in accord with their charters, these bodies have advised the government on many policy-related matters, although they are legally separate from the government.

Scientists' Organizations

Scientists' voluntary organizations, formed outside the official link between science and the government, often are quite critical of some aspect of government policy or performance, if not of the state of society in general. Nonetheless, there have been intermediate forms and links between the official and unofficial structures.

The period under study here—the period since about 1945—has seen a proliferation of voluntary public-interest organizations representing a wide variety of constituencies, which is fully in line with Tocqueville's observations. Those with similar and allied goals can be understood as parts of larger social movements. Scientists and engineers, even by a very generous definition, constitute only 0.7 percent of the US labor force (as of 1993; see NSB 1998: A-117), and scientists' organizations constitute only a small segment of the wider population of what sociologists call *social-movement organizations*. Especially when the movements grow large, as the peace and environmental movements did, organizations of non-scientists—ranging from Musicians Against Nuclear Arms to Greenpeace—tend to outnumber organizations of scientists, just as non-scientists outnumber scientists among the individual followers. When hundreds of thousands of citizens marched against the nuclear arms race, the percentage of scientists among them was probably quite small. Scientists, of course, also often participate in social movements as members of organizations not specifically geared to a scientific constituency. Conversely, numerous scientists' organizations have accepted members and supporters who are not scientists. Our inquiry is restricted to social-movement organizations of scientists and social-movement organizations in which scientists have played a dominant role; however, we shall also briefly visit organizations in the gray areas between science and other professions, including medicine and law.

Public-Interest Organizations

The social and environmental impacts of science and technology, as well as their social uses (and potential misuses), have been major concerns of what we call *scientists' public-interest organizations*. Organizations of this kind are central to this chapter. We pay less attention to the perhaps better-known and often long-established scientists' organizations that have devoted themselves to the progress of science in general (e.g., the American Association for the Advancement of Science) or to the progress of a specific field (e.g., the American Physical Society). The primary purpose of these professional organizations is internal to science, although they also look out for their members' career interests (that is, they have trade-union-like functions) and although they include some lively public-interest sections and forums.[3] In fact, we shall discuss the stances these associations have taken toward political and public-interest issues, their often complex relationships with scientists' public-interest organizations, and their own organizational subdivisions that deal with political and public-interest questions.

Special cases are the organizations that promote the interests of particular categories of scientists, such as women and minorities. Among them are the Association for Women in Science, the Women in Engineering Programs and Advocates Network, the American Indian Science and Engineering Society, the National Action Council for Minorities in Engineering, the National Society of Black Engineers, the NIH Black Scientists Association, the Society for Advancement of Chicanos and Native Americans in Science, and the Society of Hispanic Professional Engineers. Their efforts to improve the lot of previously disenfranchised and oppressed groups and to combat gender and racial discrimination can be regarded as serving the public interest, because these endeavors coincide with fundamental aspirations of American society at large.

Our focus on scientists' voluntary associations leaves for later, or to others, biography-type studies of the civic activities and influence of individual scientists. Some individuals, such as Leo Szilard, were in the forefront of organizing several of the scientists' associations (Szilard 1978; Lanouette 1992). Others preferred to act on their own, yet their influence in many cases was enormous. We are thinking here of Albert Einstein and Bertrand Russell (who collaborated on a 1955 manifesto), and of Hans Bethe, Sidney Drell, Frank von Hippel, Jerome Wiesner, Paul Doty, Philip Morrison, Bernard Feld, and R. L. Garwin, among many others. The biologist Matthew Meselson reached out privately to lawyers, who helped him draft proposed international treaties. Meselson also briefed the Canadian scientist John Polanyi that the use by the US Armed Forces of Agent Orange and tear gas during the Vietnam War might well be in contravention of the Geneva Protocol. Polanyi, in turn, persuaded the Canadian government to endorse that interpretation publicly.

Many of the actions of individuals such as those named above were so far from the wider public's attention that their contributions were not clearly visible. Sometimes their actions were taken for granted. Some of Linus Pauling's numerous undertakings were highly popularized but were more ad hoc, and typically these were not meant to, nor did they, translate into durable organizations. A major example of the activities of Pauling (who received the 1962 Nobel Prize for peace in addition to the 1954 Nobel Prize for chemistry) was his campaign against fallout from nuclear tests. Other scientists organized ad hoc letter-writing or speaking campaigns on behalf of victims of human-rights abuses, notably in China and the Soviet Union.

And of course no list of US scientists who changed history is complete without Rachel Carson, the pioneer of environmental concerns. Some of these individual energizers have already been studied, but there is more to be done.

The Waves of Citizen-Scientists' Activism

The Tocquevillian perspective draws attention to the fact that the Constitution of the United States, with its ample guarantees of civil liberties, is exceptionally conducive to a civil society with all sorts of voluntary associations. Cross-national comparisons (e.g., McAdam, McCarthy, and Zald 1996b) have focused on this often-neglected dimension. Yet, within the essentially constant constitutional framework, changing political, economic, social, and cultural factors contribute to marked fluctuations in associative activity. Rather than being randomly distributed over time, social-movement organizations appear grouped together in distinct clusters, or waves, which constitutional factors cannot be invoked to explain. The notion that social-movement activity is cyclical or oscillating is widely used and empirically documented (e.g., Klapp 1972; McCarthy and Zald 1977; Shorter and Tilly 1974; Tilly 1978). It certainly also fits the observed wave-like nature in the intensity of scientists' organizing activity.

In the post-World War II era, roughly three periods of high activity by scientists' organizations can be discerned, the first in the immediate postwar years.

The First Wave

The Manhattan Project was the most important cradle of first-wave activism. The scientists who worked on it had been recruited to try to develop nuclear weapons, initially out of the reasonable fear that Germany had a head start on the United States. But a concern analogous to that of Cincinnatus arose among the citizen-scientists associated with the project. They recognized a pressing need (stemming from the circumstances of victory, not from the menace of military disaster), and they thought that their civic engagement was required to prevent great potential harm, especially since many among them felt that their work could have contributed to it. They acted from a sense of emergency, and they did not intend to make their activist interventions a career; they would much rather have returned to their laboratories and their scientific pursuits at the first possible opportunity.

Some of the Manhattan Project scientists were from early on very interested in the social and political ramifications of what they had been called upon to accomplish. Of course, the task given to them was to build nuclear weapons as quickly as possible; but individuals of diverse convictions, including Niels Bohr, Vannevar Bush, and a group of scientists at the Metallurgical Laboratory on the University of Chicago campus, began considering the aftermath, even while the war and the Manhattan Project were still in full swing. On November 18, 1944, a committee that had been formed in mid 1944 at the Chicago "Met Lab" under Zay Jeffries to study the issue submitted a report titled Prospectus on Nucleonics. This was a precursor of the more famous Franck Report of June 1945 (signed by D. J. Hughes, J. J. Nickson, E. Rabinowitch, G. T. Seaborg, J. C. Stearns, and L. Szilard, and by the committee's chairman, James Franck), and of the petition, organized by Szilard and submitted in July 1945, in which 68 signatories asked the president to give Japan a chance to surrender before atomic bombs would be used and further urged him to consider the moral implications in his decision about using the nuclear weapon (Bundy 1988; Rhodes 1986; Smith 1965; Szilard 1978: 211).

In such ways, a number of Manhattan Project scientists, who had responded to the government's call to set the creation of the atomic bomb in motion, developed an escalating sense of concern as the project sped toward its goal. Some of the leading scientists were also the leading dissenters, and Leo Szilard was one of the most vocal. At this early stage, one might say, the two incipient ivory bridges had not yet separated. There was just one very new and unusual bridge, its builders wavering to various degrees between collaboration and dissent with respect to the government. Few people had a clear idea about standards of behavior in this novel situation. The scientists worked for the military, but their expert skills and their status set them apart from the usual command hierarchy of the armed forces. They counted on their prestige with the upper echelons of the military and the government, and their initial impulse was to influence the government directly by means of reports, petitions, and the like.

But after two nuclear bombs had been dropped on Japan, the first scientists' organizations sprung up, and thus the second connection began to acquire its separate contours. Alice K. Smith (1965: 97), the pioneering historian of this epoch, put it as follows: "The impulse of Project scientists to organize in order to educate themselves and others in the implications of atomic energy was rapidly reinforced by uneasiness about channels of communication with those who were formulating national policy. . . . The fear spread that they [the scientists] might be left without any means of voicing their opinions except such as was allowed them by the army." Leo Szilard pointed out that the atomic scientists amounted to the entirety of the democratic constituency concerning decision making about nuclear weapons, insofar as—owing to the secrecy imposed on the Manhattan Project—nobody else knew much about it.

The Association of Los Alamos Scientists was formed on August 30, 1945 (Smith 1965: 115). Concerns about the future uses of nuclear energy were also being articulated at the other major sites of the Manhattan Project. In the Clinton Laboratories at Oak Ridge, the Association of Oak Ridge Scientists came into existence in September 1945. In Chicago, another early hotbed of activity, a group called the Atomic Scientists of Chicago issued their first press release on October 4, 1945, claiming the support of 95 percent of the Manhattan Project scientists in Chicago (Smith 1965: 87). The group emphasized that with the advent of atomic bombs international control of nuclear energy was the only chance for successful stabilization of international relations, and that it was the responsibility of the atomic scientists to explain the significance of the bomb. Chicago was also the place where Hyman Goldsmith and Eugene Rabinowitch started the *Bulletin of the Atomic Scientists of Chicago*. Publishing its first issue on December 10, 1945, the *Bulletin* was to become the most important voice of the whole movement, soon dropping "of Chicago" from its title. Similar associations were started at other Manhattan Project sites, and at scientific installations that had worked on other war-related projects (notably the Radiation Laboratory in Cambridge, Massachusetts).

The specific focus of much of the first-wave activity was opposition to a War Department-sponsored congressional bill of 1945 that the scientists feared would bring nuclear energy under a strong measure of military control (Smith 1965: 128–173; Wang 1999). The citizen-scientists lobbied for a civilian-run Atomic Energy Commission, and in the end they were successful. However, farther-reaching ideas about the need for international control of atomic energy and about restricting the technology to non-military purposes were not realized.

The local organizations of scientists sprang up relatively independently, in part because communication between the sites was curtailed by the secrecy that shrouded the Manhattan Project. But of course there was some contact, and the local groups immediately undertook efforts to amalgamate and speak with a united voice. They combined in the Federation of Atomic Scientists, which was quickly superseded by the Federation of American Scientists, founded in December 1945. This second national organization had tentatively been called the Federation of Scientific Organizations, then the American Federation of Scientists, before the final name was arrived at (Smith 1965: 229, 235–236).

The organizational meeting of the Federation of American Scientists (held on December 7 and 8, 1945, at George Washington University in Washington, DC) included delegates from twelve local groups: the Association of Los Alamos Scientists, the Association of New York Scientists, the Atomic Scientists of Chicago, the Science Society of Washington, the Association of Philadelphia Scientists, the Association of Cambridge Scientists, the Rocket Research Group of the Allegany Ballistics Laboratory, the Dayton Association of Atomic Scientists, the Atomic Engineers of Oak Ridge, the Association of Oak Ridge Scientists, the Manhattan Project Scientists at New York, and a group from Rochester that was simply listed as "Rochester Section, Federation of American Scientists" (Smith 1965: 236).[4] It is perhaps noteworthy that Oak Ridge had separate organizations for scientists and engineers,[5] and New York also was represented twice, because the Manhattan Project scientists and the other scientists had distinct associations. The membership of all these groups added up to nearly 3,000 citizen-scientists (Smith 1965: 235).

Founded in 1938, the American Association of Scientific Workers (AAScW) still existed at the end of the war, and it tried to involve itself in the debate about nuclear weapons and nuclear energy. Owing to its far-left politics, it had little credibility with most of the atomic scientists or with most members of the public, and it was generally ignored. "AAScW influence in the atomic scientists organization was, so far as one can determine, trivial. A deliberate decision was made to dissociate the two movements, and the AAScW sympathizers were effectively quarantined within the Scientists Movement." (Strickland 1968: 15) This was a time for activism of the elite. Having built the bomb that had ended World War II, the atomic scientists had enormous credit when speaking out about nuclear issues. But the change from "Federation of Atomic Scientists" to "Federation of American Scientists" signaled that the latter organization's founders were aiming at the broader constituency of all scientists rather than only at the tiny elite of Manhattan Project participants.

Albert Einstein, though he had not participated in the Manhattan Project, was particularly active in the postwar movement. In October 1945, for example, he published a declaration, with the novelist Thomas Mann and a dozen prominent Americans, that advocated a global legal order as a means for preventing nuclear war (Fölsing 1993: 722). In 1946, the Emergency Committee of Atomic Scientists (ECAS) was set up, with Einstein as chairman and with a number of eminent scientists (among them Szilard and Pauling) as members. ECAS used the great prestige of the participating scientists to raise funds for the FAS and for related causes, including the *Bulletin of the Atomic Scientists*. In June 1947, a public statement by ECAS about the uselessness of the United Nations Atomic Energy Commission and about the necessity of world government contradicted the positions of the FAS and antagonized its leadership. In the wake of this disagreement, the activities of ECAS soon tapered off. In 1948, as the first wave of activism was ending, the Society for Social Responsibility in Science (SSRS) was founded. Its purpose was to get scientists to sign a pledge that they would refuse to carry out any research with direct military implications.

Between the first surge of associational activity in the immediate postwar years and the next one, a lull occurred as a result of the stifling political climate of what came to be known as the McCarthy era. This extraordinary period, during which honorable political activity could become a liability with potentially devastating effects on one's career and life, illustrates how sensitive citizens' activism can be to shifts in political parameters (Wang 1999).

The Second Wave

The second wave of activism reached its high point in the late 1960s and the early 1970s. Whereas the causes of the first wave could be pinpointed with some accuracy to developments within science itself (i.e., research in nuclear physics and technology), much wider societal factors played a large role in the expansion, if not so much in the origins, of the second wave. Now the scientists' associations formed parts of much larger social movements for peace and environmental protection. On the whole, the issues were more diffuse and much more loosely connected to science, and the scientists had a less prominent place within the larger social movements of the second wave.

The primary political issues of the time were civil rights and the Vietnam War. Yet, by themselves, these political issues may not sufficiently account for the increased social-movement activism. An additional impetus came from deeper socio-cultural and value changes. Substantial parts of the younger cohorts have been said to have adopted "post-materialistic" values, such as self-actualization, that belong to the higher levels of a Maslowian hierarchy of values (Inglehart 1977). The second wave of scientists' activism may thus be linked to those wider phenomena that expressed themselves in the emerging "counterculture," with its fundamental rejection of certain cultural standards. The citizen-scientists—even many of the "radical" ones—tended to be at the moderate end of the overall countercultural spectrum; as long as they continued working as scientists, they stayed connected to traditional standards and practices of objectivity and rationality.

"The late 1960s" has become a ubiquitous shorthand for a host of momentous changes in multiple facets of American society. With good reason, Kelly Moore (1993: 85–86) argued that a new wave of scientists' activism started in 1969, and that it was very different from its predecessors. Yet beneath the obvious breaks, important continuities connected the crest of the late 1960s and the early 1970s back to beginnings in the early 1960s and even the late 1950s. At the level of issues, an important thread of continuity was the civil rights struggle in the United States, which became prominent in the early 1960s and remained a focus of attention in the late 1960s. Furthermore, perceptive observers could see a new social group of "new radicals" (Jacobs and Landau 1966) emerge already in the early 1960s. In addition, a number of major scientists' organizations came into existence already in those earlier years. Therefore, while acknowledging that considerable changes indeed occurred in the late 1960s, we shall speak of a single and long-lasting second wave that began in the early 1960s. (Of course, our periodization, like most, is to a considerable extent arbitrary.)

The point to be made in giving the following examples of scientists' associations that sprang up in the late 1950s and the early 1960s is that they—along with the FAS, which had survived from the first wave—were already in place when scientists' activism began to boom in the late 1960s. Not only were they all lifted by that rising tide and incorporated some of the newer concerns in the process; but they also served to channel and structure the new burst of activity.

The Pugwash Conferences were initiated in 1957, partly in response to the Russell-Einstein manifesto, which Einstein had signed two days before his death in 1955. The initial focus of Pugwash was on involving experts from the United States, the Soviet Union, and many other countries in a search for political strategies for preventing nuclear war, but its growing agenda contained from the beginning also a concern for the hazards of the peaceful use of nuclear energy. In its organization as well as its roots, the Pugwash Conferences (bringing together an international group of well-known scientists and other public figures) initially resembled the first-wave pattern of focusing on elites. The later addition in 1979 of a vigorous "Student Pugwash" wing reflected the changing times of a younger and more rank-and-file-based activism. (In 1999 there were eighteen national Student Pugwash organizations around the globe, including an energetic group in the United States.)

Another one of those early organizations whose activities continued through the crest of the second wave was the Committee for a Sane Nuclear Policy (SANE)—renamed "SANE/Freeze" in 1987 after merging with the Freeze Campaign, and renamed again "Peace Action" in 1993. SANE was founded in 1957 in response to Albert Schweitzer's "Call to Conscience," with the goal of promoting global nuclear disarmament. Although not a scientist-originated organization, it counted many scientists among its members (Nichols 1974: 134). In the 1960s and the 1970s, SANE expanded its mission, joining the anti-Vietnam War movement and forming alliances with organized labor.

In 1962, Leo Szilard founded the Council for a Livable World. Dedicated to combating the danger of nuclear war, it has continued to operate as a political lobby and a political action committee, supporting candidates for political office who are sympathetic to its causes.

No listing of these early associations would be complete without mentioning the Scientists' Institute for Public Information (SIPI) and its forerunners. In the late 1950s and the early 1960s, concern was rising in the St. Louis area about the health hazard of radioactive fallout, especially after the development and testing of hydrogen fusion bombs produced fallout that could circulate more widely in the atmosphere than that of fission bombs could. St. Louis became the birthplace of several citizen-scientists' associations that, from 1958 on, addressed the fallout problem. Barry Commoner, a biologist at Washington University, was the leading activist of this movement.

The fallout issue can be seen as one of the links between anti-nuclear-war activism and environmental concerns. As ecological questions moved to the fore, even the *peaceful* use of nuclear energy eventually became subject to skeptical scrutiny. Many of the first-wave citizen-scientists had been gratified when President Eisenhower promoted "Atoms for Peace" in a famous speech on December 8, 1953, before the General Assembly of the United Nations (Weart 1988: 158), and when two international conferences on that concept were held. But now the focus shifted onto the hazards of civilian uses of nuclear power (Nelkin 1971). Within an expanding environmental agenda, the nuclear issue—much more widely defined than just nuclear war—combined with general concerns about other environmental dangers of modern technology. Rachel Carson's acclaimed 1962 book *Silent Spring* had become a major energizer of a "green" movement whose scope encompassed multiple degradations of the environment. The pollution of American rivers was shockingly highlighted in 1969 when the Cuyahoga River in Cleveland caught fire. This was a wake-up call for many people. In 1970, the first "Earth Day" provided a spectacular rallying point for the fledgling environmental movement.[6]

The crest of the second wave in the late 1960s saw an explosion-like multiplication of scientists' public-interest organizations. Between 1968 and 1970, at least 35 new organizations were founded (Moore 1993: 96). A highly significant event occurred on March 4, 1969, when thousands of scientists held a one-day research stoppage to discuss and protest the close link between academic research institutions and the military, and especially the use of science-based technologies in the Vietnam War (Allen 1970; Leslie 1993; Nelkin 1972). The work stoppage was initiated at the Massachusetts Institute of Technology, but scientists at other institutions also participated. The Union of Concerned Scientists originated in this collective action.

A more radical wing of citizen-scientists added confrontational "direct action" to the repertoire of activism. Direct action consisted largely of local protests and demonstrations. At the national level, the disruptive and theatrical tactics of the (predominantly young) group called Scientists and Engineers for Social and Political Action (SESPA)—later called Science for the People (SftP)—at conferences, such as the annual meetings of the American Association for the Advancement of Science 1969 in Boston and 1970 in Chicago, grabbed some attention. In Chicago, for instance, Edward Teller, who was remembered for his advocacy of the H-bomb, was presented a "Dr. Strangelove" award (which he refused), and the new AAAS president who also was Atomic Energy Commissioner, Glenn T. Seaborg, was threatened with a mock indictment for "the crime of science against the people." Furthermore, there were political workshops, guerrilla theater, and a pledge of non-participation in "war research" (Nichols 1974: 157, 170).

Relative to the much more violent and serious tactics that some other social-movement organizations employed during that era, the activities of even the most "radical" scientists or students were relatively tame. Nonetheless, the revolutionary rhetoric was apt to shock or even scare the scientific establishment, as was intended. For instance, a SESPA activist, in an unscheduled speech at the 1970 AAAS meeting (cited in Nichols 1974: 156), said:

> What is needed now is not liberal reform . . . but a radical attack. . . . Scientific workers must develop ways to put their skills at the service of the people. . . . You still have the opportunity to work constructively with the movement for revolutionary change. . . . But if no other solution is available, we will be out in the streets, . . . doing everything we can to tear this racist, imperialist system to shreds. . . .

In New York, another radical group, Computer People for Peace, forged an alliance with the Black Panthers, for whom they collected bail money. In the field of medicine, the Medical Committee for Human Rights, founded in 1964, could be considered the counterpart of SESPA/SftP in terms of its radical goal of attacking the capitalist system as such. The moderate scientists' organizations, however, pursued much more limited goals, such as protesting the building of an anti-ballistic-missile system (Brennan et al. 1969; Chayes and Wiesner 1969) or a supersonic airplane (Shurcliff 1970) and calling for the establishment of safety codes against environmental hazards. They did so through conventional channels of influencing the public and the policy makers. At the international level, a major focus for the moderates was the protection of colleagues' human and scientific rights worldwide, and especially in what at the time was the Communist Bloc. In 1972, the Committee of Concerned Scientists was founded to help oppressed scientists.

The societal aspects of science also increasingly became a topic of academic research and teaching. Starting in 1966, an informal Science and Public Policy Study Group of the AAAS became instrumental in helping academic programs on science and technology policy off the ground. In Britain, Science in a Social Context was founded in Leeds in 1972 to foster a greater emphasis on the social aspects of science in science instruction at the undergraduate level. Both in Britain and in the United States, programs in "science studies" and in "science, technology, and society" sprang up, aided by these initiatives. In 1975, the Society for Social Studies of Science was founded.

The study and teaching of the history and of the sociology of science had already been in existence as professional activities for decades. One thinks of pioneers such as George Sarton and Robert K. Merton. Yet in the early 1970s both disciplines, and the new discipline of "science studies," experienced intensified interest and growth. Among some sociologists of science, this was accompanied by a shift in focus from the social structures in which science as a social activity was performed, to the social determinants of scientific knowledge itself. Analogous developments occurred in the history of science. Many academics studying the history and sociology of science brought their own perspectives to bear on the two intersections of science and society—perspectives that often differed markedly from those of the scientist-administrators and citizen-scientists themselves. They tried to redefine the meanings of both the ivory tower and the ivory bridges that lead from it into society, posing troubling questions about the nature of science and its relationship to society.

The Third Wave

In the early 1980s, after the second wave had receded somewhat in the latter part of the 1970s, a new wave appeared, to some extent in response to an intensification of the Cold War and to a nuclear arms buildup that had begun under President Carter (Casper and Krauss 1981). Perhaps also in an attempt to counteract the growing wave of activism, President Reagan announced the Strategic Defense Initiative (labeled "Star Wars" by its opponents)—a project to develop a defensive system that would heavily rely on hoped-for advances in science and technology. The reaction to SDI was mostly negative among the citizen-scientists, who renewed their interest in the peace issue without diminishing their attention to environmental issues. The continuing emphasis on the environmental risks of modern technology had been heightened in the late 1970s by alarming and highly publicized events such as the 1978 environmental disaster at Love Canal and the 1979 nuclear power plant incident at Three Mile Island. Other catalysts included oil spills from wrecked supertankers that devastated coastlines and marine life. More recently, concerns about climate change, global warming, and the degradation of the biosphere have come to the fore.

It is open to debate whether we are currently in a continuation of the third wave or in a fourth wave. Both continuities and shifts can be discerned since the 1980s. In the following we shall speak of one long third wave in which the structure of the scientists' organizations has remained somewhat constant but the agenda has become enlarged. With the end of the Cold War around 1990, one of the major goals of scientists' activism was achieved (although it not certain how strong an impact the scientists' associations had on this turn of events). Yet new issues have quickly arisen in the changing landscape of the politics of peace and disarmament, such as nuclear non-proliferation, conversion of the military apparatus in the former Eastern Bloc, and land mines.

A third major issue has begun to coalesce that may eventually grow to rival the traditional issues of peace and the environment. It concerns the multifaceted societal implications—social, ethical, legal, economic, and cultural—of rapid advances in biotechnology and genetics. Vigorous research science is clearly and immediately the driving force behind these developments, and the role of biologists, biochemists, and geneticists in any social movement that might form around this issue can be expected to be as crucial as the role of nuclear physicists was in the first wave. One scientists' public-interest association that has expressly devoted itself to this issue is the Council for Responsible Genetics, founded in 1983.[7] We expect that genetic and related research areas will become prominent concerns for many scientists' voluntary public-interest associations. Significantly, the Human Genome Project, an important example of the first ivory bridge in this emerging area of genetics and biotechnology, established, at its inception in 1990, an Ethical, Legal, and Social Issues working group, which has been receiving 3–5 percent of the total annual project funds for the purpose of exploring the social implications of its mission and its eventual technical successes.

An emerging fourth issue of potentially major importance is information technology (with global electronic communication networks, momentous advances in computing technology, and the rapid popularization of computers). Citizen-scientists have begun to form associations in this area, such as Computer Professionals for Social Responsibility and the Center for Democracy and Technology.

The transition between the second and third waves was much less distinct than that between the first and second waves. A large number of the second-wave activists, who had embarked on activism in their youth, were still around for the third wave. In the meantime many of them had advanced in their scientific careers and had gained expertise and seniority that lent added weight to their views, both within and outside the scientific community. That now middle-aged cohort of former young activists could provide continuity, expertise, and leadership to the third wave.

Scientists' associations have become strongly entrenched in the regular political process by testifying before Congress, by lobbying, and by acting as political pressure groups. Many associations have taken advantage of the US Tax Code and formed educational divisions that qualify for tax-exempt status and are permitted to accept tax-deductible donations. Public relations and fund raising have largely been put in the hands of professionals, superseding the amateurish initial efforts of scientists in these domains. In general, the scientists' public-interest associations now have paid professional staffs in their headquarters (which are typically located in Washington). The link to local members is often maintained through chapters. The tendency toward professionalization began as the second wave was nearing its end. The Union of Concerned Scientists retained a public-relations firm in 1974, and soon thereafter it hired a full-time public-relations person (Moore 1993: 262). In the third wave, however, the use of professional public relations has become typical.

Revolutionary advances in electronic communications may have dramatic effects on the third wave and on voluntary public-interest associations in general. Scientific and technical progress in this area is not just a topic of citizen-scientists' activism; it also provides a vehicle for it. Innovations such as the Internet will be of great significance for the future of social movements. Social-movement organizations (for which the term *non-governmental organization*, abbreviated NGO, is in wide use) increased in popularity in the 1990s, at least partly as a result of the instant and easily available world-wide communication that the new information technology provided. About 6,000 international NGOs (in all fields of interest) existed in 1990; by 1999 there were more than 26,000 (*Economist*, December 11, 1999, p. 20). NGOs have also proliferated at the national level.

The Internet allows geographically scattered individuals with similar interests to unite in an immediate and inexpensive way, and it provides associations with an effective method of addressing and informing their constituencies. Scientists are one of the most computer-literate and Internet-literate groups in the population, and today there is rarely a scientists' public-interest organization that does not use the Internet to alert its constituents to emerging issues and causes.

The Citizen-Scientists: Who, Why, and How

Membership

It is relatively easy to describe the group of first-wave activists. As nuclear weapons dominated the agenda of scientists' activism, physicists were predominant among the activists. Many citizen-scientists, as Manhattan Project luminaries, were scientific leaders whose prestige, both with the federal government and in the general public, derived from their vital contributions to the victorious war effort. The public esteem, while tending to focus on nuclear physicists, accrued also to other scientists who likewise had achieved crucial breakthroughs in science and technology (e.g., radar and synthetic rubber) and in medicine. Some of these also participated in activism. In general, the first-wave citizen-scientists tended to come from the elite of the scientific community of the time.

The second wave gave more prominence to younger and junior members of the profession, such as postdoctoral fellows and assistant professors. There was also a stronger element of student activism. Thus the leaders of the citizen-scientists became more distinct from the scientific leaders. However, the leading citizen-scientists still tended to be associated with institutions of great scientific renown, including MIT, the University of California at Berkeley, the University of Wisconsin at Madison, Columbia University, and Harvard University.

As the agendas of the various scientists' public-interest associations became more varied, so did their disciplinary compositions. The relative predominance of the physicists waned. Biologists became more important, especially because they were close to the rising environmental and "green" concerns. For example, most members of the Scientists' Institute for Public Information were biologists (Moore 1996: 1605). Today we find a great variety of disciplinary backgrounds and academic seniority among the citizen-scientists, with biologists in a strong position.

Reasons for Joining

It is no deep sociological mystery why there are usually enough people to fill the available positions for scientist-administrators. The sociologist Max

Weber famously distinguished three basic dimensions of social position: wealth, prestige, and power. Scientist-administrators tend to do fairly well on all three dimensions, but the most palpable perquisite of their job probably lies in the power dimension. Furthermore, significant rewards in terms of professional prestige may accrue for insider service, and the scientist-administrators' home institutions are usually supportive of their stints in Washington. The rewards in social position for "outside" activism are less immediate and less certain. Hence, the citizen-scientists' motivation deserves a more extensive treatment.

The common rhetoric of activists in general contends that their activism follows immediately, and almost "naturally," from certain severe grievances. Yet although those certainly play a crucial role, various additional conditions and causes obviously must be present if problems and grievances are to be translated into social activism. Several explanations have been advanced for the rise of social movements, but none has been entirely satisfactory (a situation that is not unusual in the social sciences). We turn first to broad societal factors. Socio-economic deprivation (absolute or relative) may be thought to be at the root of people's activism. In this view, economic, social, and political crises create strains and dissatisfactions that lead to social movements. However, that explanation does not apply as generally as one might think. Research in this area has found it more likely that prosperity breeds associational activity. McAdam, McCarthy, and Zald (1988: 702) put it this way: "Wealthy societies may create the opportunities for entrepreneurs of grievances to attempt to develop new social movement products." In line with this general observation, the development of scientists' activism has not been driven primarily by political, social, or economic deprivation. Certainly scientists were not a marginal group whose plight was to be alleviated by a scientists' social movement. On the basis of education, occupational prestige, and income, scientists could, on average, be counted among the middle class, if not the upper middle class, and science students had a reasonable chance of achieving that privileged status. Significantly, the epicenters of scientists' organizations were at the more prominent universities, whose students and faculty had above-average career prospects.

The majority of scientists' organizations has indeed professed a concern for the public good as their leading motive. For them, the key word has been *responsibility* rather than *rights* (the latter being the central notion for numerous other interest groups). Citizen-scientists, on the whole, have viewed themselves as guardians of the public interest who promote the socially responsible use of science and technology. They share this general goal—while often disagreeing on concrete projects—with the scientist-administrators on the other connector between science and society. Similarly, two larger social movements with which many scientists' associations have been allied, the peace movement and the environmental movement, have portrayed themselves as championing universal concerns that affect all members of the population.

Still, a cynic might dismiss this lofty rhetoric as an ideological smoke screen and try to interpret scientists' associations as vehicles for the already privileged to get even more advantages. Yet this would make little sense. Under the widely prevailing reward structure for scientific careers—with research publications as the coin of the realm—careerist self-interest seems highly implausible as the fundamental motivation for individual scientists to engage in civic activism. Most of the citizen-scientists with distinguished science careers achieved their professional success independent of—in some cases perhaps even in spite of—their activism (e.g., Linus Pauling; see Goertzel and Goertzel 1995).[8]

Psychological explanations of an individual's engagement in activism that emphasize personality traits—alienation, deprivation, and similar individual psychological characteristics—have had rather limited success. Likewise, no strong links could be found between individuals' attitudes and their participation in social movements (McPhail 1971). More specific, whereas it seems rather obvious that an individual's hostility toward a certain movement prevents him or her from joining it, sympathetic attitudes are certainly not sufficient explanations for joining. As a rule, only a small proportion of those who share an attitude or goal actively participate in a corresponding social movement. The real question, then, is "Why do certain individuals of a given attitude become activists when the majority do not?"[9]

In addition to considering broad socio-economic factors and individuals' traits and attitudes, it may be useful also to examine the immediate social environment in which people live and interact.[10] Explanations of social movement participation that focus on the social environment point to critical mass phenomena (a concept of nuclear physics that can be usefully applied to the social domain) and positive feedback loops (which are similar to what Merton, in a different context, termed the Matthew Effect).

According to studies in movement formation, well-connected individuals tend to be prominent in joining and promoting new movements (Lipset

1950, see Pinard 1975). Thus, an individual who already belongs to one association is likely to join other associations (Orum 1972, McAdam 1986, Barnes and Kaase 1979, Walsh and Warland 1983). "Joiners" are apt to keep on joining, not the least because they become identified as prime targets for organizers.

A concentration of like-minded people has been identified as a crucial pre-condition for organizing. Informal communication networks are "the basic building blocks of social movements" (McAdam, McCarthy and Zald 1988:711). Informal groups mobilize members, wider communication networks, and leaders. They are also the sites for what has been termed "frame alignment" (Snow et al. 1986, Gamson 1992), in which new worldviews are collectively constructed. Paramount here is the creation of new forms of meaning, of a collective identity, and of a shared sense of purpose, not the gathering of new observations or empirical evidence.

It is indeed a major finding of the sociology of social movements that pre-existing clusters of solidarity and social infrastructures are essential for the emergence and the growth of social movements (Evans 1980; Freeman 1975; Morris 1984; McAdam 1982; McCarthy 1987; Wilson and Orum 1976; Useem 1980). Although no exhaustive explanation exists of why persons become citizen-scientists, or activists of any kind, one of the most powerful predictors of an individual's decision to join a social movement has been identified as prior contact with a participant in the same movement (e.g., McAdam 1986; Gerlach and Hine 1970).

In view of these general findings, one would expect the communities of scientists at the various Manhattan Project sites to have had a strong impact on the creation of scientists' public-interest associations. As we described in detail earlier, these local groups of scientists were indeed critical in initiating the first wave of citizen-scientists' activism. In addition, many members of what was then a relatively small national physics community knew one another personally and thus had an informal communication network that could be used to promote the citizen-scientists' associations.

In examining the emergence of second-wave scientists' associations, it appears essential to consider the effects of the agglomeration of students, graduate students, and postdoctoral fellows on campuses; the fact that universities provide an organizational framework for the interchange between practitioners of diverse disciplines; and perhaps also the urban stimuli that metropolitan areas provide. First of all, university campuses were major breeding grounds for the counterculture in the late 1960s, and they exposed the young scientists to the leaders of this phenomenon. "Organizational intellectuals" (Zald and McCarthy 1975), who have been described as a "new class" (Brint 1984; Bruce-Briggs 1981), held forth and generated countermeanings to official beliefs, which in turn stimulated the larger social movements of that era. Close proximity to these "organizational intellectuals" on campuses may have contributed to some of the scientists' social activism. This suggestion is supported by the observation that various strands of activism tend to cluster and cross-fertilize. It is certainly noteworthy that Berkeley, California, and Cambridge, Massachusetts—university towns considered to be, from time to time, hotbeds for activism in general—became extraordinarily prolific in spawning scientists' public-interest organizations.

It was found that dramatic and highly publicized events can serve as triggers or crystallization points for activism (Walsh 1981; see Useem 1980). Dramatic events were certainly also important in the origins of the scientists' groups. The nuclear destruction of Hiroshima and Nagasaki was a highly significant and evocative incident. Shortly thereafter, the legislative controversy about what kind of agency should control the further development and use of atomic energy provided a concrete focus for first-wave activism regarding nuclear policy. The 1969 work stoppage at MIT was another concrete event that caused a group of citizen-scientists to coalesce. Various environmental disasters and the rallying point of "Earth Day" had similar effects for the environmental movements. Organizations can transform such spontaneous activism and concrete goals into long-term strategies.

Types of Scientists' Associations

Among social movements, scholars have distinguished between "conflict movements" and "consensus movements" (Schwartz and Paul 1992; Lofland 1989; McCarthy and Wolfson 1992), or between "challengers" and "polity members" (Lo 1992). The former tend to engage in protest politics, the latter to practice more conventional politics (Lo 1992). William Gamson (1968) emphasized that groups relatively short on power and resources tend to use more radical strategies. Having fewer conventional avenues of influence, such groups have less to lose by opposing the established power structure.[11] All this seems applicable to the useful distinction between "radical" and "moderate" scientists' groups proposed by Nichols

(1974), the radicals corresponding to conflict movements with unconventional strategies and the moderates to consensus movements with conventional strategies. This distinction should, however, not be understood as a dichotomy between two sharply distinguished and internally homogeneous sets of organizations, but as an ideal-typical definition of the endpoints of a continuous spectrum.

Moderates and Radicals

Both kinds of scientists' groups have shared a dissatisfaction with public policy as related to science, but they have also differed in a number of ways.

First, the goals were dissimilar between moderates and radicals. According to Nichols (1974: 148), the moderate groups displayed "a typically American pragmatism" and advocated piecemeal reform. The radical groups wanted farther-reaching, even revolutionary changes that involved attacking "the system" or (more specific) the "military-industrial complex."[12]

They differed, second, in their political strategies and actions. The moderates preferred research, education, information dissemination, and lobbying. Characteristically, these efforts included sponsoring candidates in congressional elections, influencing elected politicians, and offering expert testimony. The radicals used more confrontational tactics, such as sit-ins, conference disruptions, and other forms of direct action and civil disobedience (Nichols 1974: 130–131).[13] Increasingly, legal action is being used as a strategy by both moderate and radical groups.

Third, the differences in strategies and actions were reflected in differences in funding. The moderate organizations found it relatively easy to solicit support for their projects from private foundations, whereas organizations with a radical anti-status-quo agenda normally did not get such funds as easily (Moore 1993: 318–324).

Fourth, the moderate groups included larger numbers of older scientists and established scientific leaders, whereas the radical groups typically drew a younger and scientifically junior crowd. For example, the Union of Concerned Scientists emerged from the general unrest of citizen-scientists at MIT as a relatively moderate faculty organization; students pursuing a more radical agenda soon merged into Science for the People. The modal rank of SftP's founding members was graduate student; that of the founders of the UCS was full professor (Moore 1996: 1605).[14]

The presence of prominent scientists in public-interest organizations has considerable effects on other aspects of those organizations. These widely respected individuals provide the organizations with legitimacy and prestige that can be used for raising funds, for influencing public opinion, and for getting the attention of people in power. Such opportunities exert a strong pull toward moderate politics. As might be expected from the general typology of social movements (mentioned above), the more renowned scientists indeed charted a course of consensus politics for the organizations they dominated, whereas groups of junior scientists were more attracted to conflict politics.

The Political Spectrum

Whereas an overall balance between two approximately equally strong camps of Democrats and Republicans, or of liberals and conservatives, has characterized much of general American politics in the postwar era, such a balance has not existed among the population of scientists' public-interest associations. Here one finds a decidedly "liberal" bias, especially in the first and second waves. In the third wave, interestingly, the political spectrum of scientists' organizations has changed somewhat. During the second wave, the radical organizations usually held socialist or Marxist worldviews, and even the moderates were typically "left of center." While the wider scientists' movement lost many of its radical left-wing components, and in that sense became more moderate, there now is also a vocal minority of scientists' organizations at the conservative or "right" end of the spectrum. For example, the Science and Environmental Policy Project, headed by Fred Singer, vigorously opposes the view that the world climate is undergoing a human-made process of global warming (the "greenhouse effect") and warns against what it considers premature regulations (e.g., those proposed in the Kyoto Protocol). Another organization with similar goals is the George C. Marshall Institute.

One of the reasons for the overall liberal slant is probably that scientists who supported nuclear armament and did not champion "green" issues did not feel the need to protest the activities that occurred in government-directed science—that is, on what we call the first ivory bridge linking science and society. Instead, they typically had ample opportunity to participate in those activities. And if they wished to express their views about science and society, they could easily find other outlets in semi-official organizations or conservative think tanks. It is perhaps telling in this respect that once some goals of the environmentalist movement were institutionalized on the first ivory bridge, via the Environmental Protection Agency and numerous environmental regulations, the conservative wing of the scientists' public-interest associations began to grow.

Another reason for the "left" bias among scientists' public-interest associations might be that the group of conservative-leaning scientists was more likely than their left-leaning colleagues to interpret the ideal of scientific neutrality as demanding a measure of detachment from, rather than involvement in, the political sphere. After all, a long and strong tradition on the left, going back to Bernal, has advocated the societal usefulness of science as a primary goal, whereas conservative scientists might be more likely to identify with the "ivory tower" ethos of science.

Organizational Dynamics

Many volunteer groups are ephemeral. In the prevailing "grassroots" mode of organization building, the idiosyncratic efforts of dedicated but overworked individuals are all-important. Scientists might get together informally in their departments and laboratories, form loose groups, grope around for existing organizations of similar-minded colleagues, and perhaps raise the funds to invite a speaker or two, but then might just as quickly fall back into passivity, perhaps when the person who provided the driving force moves to another job or reorients his or her priorities to scientific work. Organizations that are too short-lived to be described are nonetheless important testing grounds for the citizen-scientists' activism. There, new issues are often experimented with, and ideas are developed that may eventually influence the better-established organizations.

When it comes to explaining at the personal level how individual citizen-scientists sustain their activism over extended periods of time, role theory makes the interesting point that a pattern of activism, perhaps begun in a phase of "biographical availability" (McAdam 1986: 70), might endure in a person's life if the social role of "activist" becomes an integral part of the person's identity. Furthermore, a career in activism implies a high amount of biographical investment (Becker 1963). That is, a habitual activist has spent a great deal of energy and time acquiring special skills and experiences that would become obsolete if he or she were to leave activism. For these reasons, a committed activist may even go looking for new a cause if the original cause is resolved.

However, just as scientist-administrators tend to resist becoming mere administrators, citizen-scientists resist becoming mere activists. Many of them eventually leave activism, drawn back to science by a strong allegiance to the self-image of being a scientist.

When organizations mature, they tend to undergo certain changes. The syndrome of bureaucratization and oligarchization is typical, although not the only possible trajectory of development. Weber classically called this process the "routinization of charisma."[15] Concomitant changes tend to occur in the organizations' goals. Organizational sociology's "law" of "goal displacement" (Merton 1968; Michels 1949; Sills 1957) seems relevant here: Once an organization is established, its own survival becomes a major objective—sometimes so important an objective that it can rival or even eclipse the organization's original purpose. Consequently, an organization may replace its original goals with more general and abstract goals that are not likely to be accomplished anytime soon.[16] Such a pattern can be documented in several of the longer-lived scientists' organizations. For example, as we have already noted, the original goal of the Scientists' Institute for Public Information was to warn of the dangers of radioactive fallout in the St. Louis area. After a treaty partially banning the testing of nuclear weapons was signed in 1963, the fallout issue declined in importance and SIPI turned its attention to broader environmental concerns (Nichols 1974: 144).

The contrast between Scientists for Sakharov, Orlov, and Shcharansky (SOS) and the Committee of Concerned Scientists—two organizations founded to help dissident scientists in the Soviet Union—is instructive. SOS stayed primarily focused on the individuals named in its title, and it folded when those scientists were released. The Committee of Concerned Scientists generalized its program to support other persecuted scientists, and it still exists today. Similarly, in the 1970s the Union of Concerned Scientists widened its agenda to include nuclear power generation. Parallel agenda shifts also occurred in other major organizations, such as the Federation of American Scientists, the Council for a Livable World, and Pugwash. These changes can be described as "diversifying the issue portfolio." Decisive goal shifts are rare. What usually happens is that some goals receive more emphasis while others fade into the background as the political climate and the societal situation change. Yet the deemphasized goals are not abandoned and may become more important again at a later point. Many scientists' organizations have now built such a diversified goal base that it is hard to imagine, if not theoretically impossible, that they would ever have to dissolve themselves for reasons of complete success.

Centralization, bureaucratization, and professionalization often go hand in hand with the changes posited by the "law" of goal displacement. A second sociological "law"—the "iron law of oligarchy" (Michels 1949)—asserts that, over time, an organization tends to become dominated by a relatively small group of people, even if volunteer efforts still occur in the chapters (or in otherwise named local units).

The establishment of a national office with full-time and/or part-time officers and a paid professional staff is a typical milestone in the development of a scientists' public-interest organization. The Scientists' Institute for Public Information has reached what is probably the final stage of professionalization. From its roots in local St. Louis fallout information groups in 1963, SIPI developed into a sought-after national disseminator of scientific information. It now is defunct as a scientists' public-interest organization; however, its highly successful Media Resource Service, instituted in 1980, lives on. MRS helps journalists locate scientists who are prepared to talk to the media on scientific topics. It now has a database of 30,000 such scientists in various fields of expertise. Since 1994, it has been run as a no-fee professional service by Sigma Xi, The Scientific Research Society (a supra-disciplinary scientific association), with funding from major foundations, corporations, and media outlets.

The Federation of American Scientists has undergone considerable changes in its long history. Founded almost immediately after World War II as a federation of local scientists' associations (most of them connected with the Manhattan Project), the FAS still exists and is recognized as one of the major scientists' public-interest associations. The year 1970, in particular, was a watershed. In that year Herbert York (who had been Director of Defense Research and Engineering in the Eisenhower administration, and who thus was a prominent "crossover" from the governmental ivory bridge) became the organization's chairman, and Jeremy Stone, who was to play a pivotal role in the further development of the FAS, became its full-time director (Nichols 1974: 135). This made the FAS, which earlier had floundered somewhat, more bureaucratized, more centralized, and more professionalized.

Science for the People, in contrast, collapsed, partly because it refused to go down this path. SftP consciously adhered to the grassroots and anti-establishment ethos of its early days. There was great resistance to its becoming a more regular organization. The traditional model of dues-paying members was rejected; membership was defined as participating in the activities of the group (Moore 1993: 97). Endless debates about the group's purpose and its structure produced paralysis. What made this grass-roots organization different from those of the earlier wave was that SftP made its organizational form a cherished principle. It engaged in introspection and self-examination to an uncommon extent, focusing on the process by which results were to be accomplished and perhaps less on the results themselves. For instance, discussing issues until a universal consensus emerged was considered very important, and that laborious process was itself considered an achievement. Monetary troubles finally caused SftP (which resisted the "iron law of oligarchy" to the end) to fold.

In general, scientists' organizations have conformed to the pattern of bureaucratization and to related trends that organizational sociologists have observed frequently when organizations age. We should, however, also note some unique features that distinguish the organizations' trajectories in different waves of citizen-scientists' activity.

During the first wave, spontaneous organizing started in local groups of scientists at the large laboratories of the Manhattan Project (and of other war-related scientific efforts). From these decentralized, "grass-roots" beginnings, national organizations such as the Federation of Atomic Scientists and later the Federation of American Scientists quickly evolved—in a classic case of a "bottom-up" process of amalgamation.[17]

Similarly, many second-wave associations started as spontaneous, decentralized volunteer efforts. In some cases, the grassroots mode of organization became a goal in itself (e.g., Science for the People), and hence bureaucratization was resisted with uncommon vigor, whereas most associations followed the more usual path of development.

Third-wave organizations tended to be somewhat more professionalized from the start. They could model themselves after a large number of successful associations and perhaps even get assistance from them, thus reaping the benefits of the organizational infrastructure that had been in place from earlier waves.

The Ecology of Scientists' Public-Interest Associations

No organization exists in a vacuum. This section addresses the various relationships of scientists' voluntary public-interest associations among themselves and with other organizations. According to our earlier definition of the scope of this study, we have been focusing on (1) voluntary (2) scientists' (3) public-interest associations. We will now explore the margins of the space thus defined, by relaxing one of these elements at a time, and examining the type of organizations that come into view.

Voluntary Organizations and Government Institutions

The scientists' public-interest organizations relate to the institutions of governmental science policy. Even after these two bridges between science and society differentiated from each other after World War II, people still moved back and forth; and some scientists were active on both simultaneously. As one might expect, the moderate citizen-scientists were more likely than the radicals to participate in first-bridge activities. For instance, among the founding members of the Union of Concerned Scientists, about 75 percent had worked for a government agency or committee, as compared to about 50 percent of the founders of the Scientists' Institute for Public Information, and only 10 percent of Science for the People's founders (Moore 1993: 163). The disparities in seniority among these three groups of course also help explain these dramatic differences in participation rates.

The height of the second wave of scientists' activism coincided with the largest degree of divergence between the two bridges. In the late 1960s, a substantial number of scientist-administrators moved outside the government and tried to influence policy from the outside, especially on the issue of anti-ballistic-missile systems, after they had lost the debate within the executive branch in 1967. Other factors in that exodus were disagreement with the Indochina policies of the day, and a feeling that the internal science advisers' influence was diminishing in the Johnson and Nixon administrations (Primack and von Hippel 1974: 262)—and finally of course the abolition of the top of the presidential science advisory structure. The vehicle of choice through which many former insiders worked was the Federation of American Scientists. By 1972, two scientists with strong former connections to the Department of Defense, Marvin Goldberger and Herbert York, had chaired the FAS.

Once the radical noises subsided and the scientists' public-interest organizations, collectively, had become more moderate, the cross-links between the two bridges were strengthened. We have already mentioned that environmental concerns have crossed over from the second to the first bridge where the Environmental Protection Agency and a host of environmental regulation have been instituted. Furthermore, the condition of women and minorities in the sciences and engineering, which has been the focus of a number of scientists voluntary associations, was recognized as a major issue by governmental science policy. The National Science Foundation, for instance, received the express mandate in 1980 to increase the participation in science and engineering of women and minorities, and persons with disabilities. A multitude of initiatives have been undertaken at various levels, from individual university departments to nationwide programs, to improve the participation and career outcomes of these hitherto underrepresented groups in the science and engineering fields.

Scientists' Organizations and Other Organizations

Relative to the wider peace and environmentalist movements, scientists' public-interest associations function as a kind of information elite or "scientific muscle." They provide relevant information not only directly to the public at large, but also—importantly—to non-scientific activist groups. These other groups can then use that scientific information to credibly counter the claims emanating from the governmental ivory bridge. The scientists' associations thus lend legitimacy to the wider movement.

There are intermediate forms between scientists' associations and other associations. Some groups, such as SANE/Freeze, have never been exclusively scientists' groups but have had strong contingents of scientists among their ranks. Others have developed a large base of non-scientist contributors or supporters. In view of the limited size of the scientific community, appeals to the general population (or to the general population of movement sympathizers) may be necessary to boost the finances of scientists' associations. Within the larger social movements, scientists' associations are often in contact with "elite" associations formed by other professionals.

Medical Organizations

Centering on people's health, the ethos of the medical profession possesses a strong and obvious affinity to altruistic ideals. Many physicians, nurses, and other professionals in the medical field have volunteered their skills in the service of health-related humanitarian causes. They have formed associations to combat major health hazards, such as those associated with nuclear war, or to provide medical assistance in disasters and emergencies. Notable examples among these organizations are International Physicians for the Prevention of Nuclear War and Médecins Sans Frontières (Doctors Without Borders). Both are international organizations, and they have received Nobel Peace Prizes (IPPNW in 1985, Médecins Sans Frontières in 1999) in recognition of their work aimed at reducing health risks and human suffering worldwide.

At the national level, the Medical Committee for Human Rights was founded in 1964 to provide medical support for participants in civil rights marches. This association of medical professionals pursued a radical agenda that included the rejection of capitalist society. In contrast, Dentists for Peace, Physicians for Social Responsibility, International Physicians for the Prevention of Nuclear War, and Médecins Sans Frontières were moderate.[18]

There is an obvious proximity between medical organizations and scientists' organizations. Many rank-and-file members and functionaries of the medical groups work at medical schools and thus may be involved in research, or in close professional and social contact with their scientific counterparts, e.g., on the university faculties. (Increasingly there are also connections in the non-academic sector, e.g., in the biotechnology industry.)

Lawyers' Organizations

Another profession whose voluntary associations for the public good often overlap with scientists' organizations is the legal profession. There is certainly also a wide array of legal public-interest organizations with no particular link to scientific issues; but within the scope of this study, we concern ourselves here only with those associations that combine scientific and legal aspects.

Two important ways for voluntary associations to influence society are political and legal activism. The two are of course not mutually exclusive, and organizations might vacillate between them, depending on which appears more promising under the given circumstances. Yet it still makes sense to distinguish between a strategy that aims at influencing the public and the politicians to create laws, regulations, or treaties in the interest of the group, and a strategy that pursues the interests of the group within the existing legal framework by challenging relevant laws or regulations, or vigorously pursuing violators of certain laws.

Both strategies can benefit from legal expertise, but especially the second one. The latter is also becoming the more promising, the more laws and regulations are already on the books. Issues of public interest are increasingly being fought over in the legal arena, which makes the services of lawyers more and more crucial. For instance, although there has also been a countercurrent (especially during the years of the Reagan presidency) of limiting or weakening environmental regulation, the last decades of the twentieth century brought an enormous increase in legislation designed to protect the environment. The legal aspect is therefore particularly prominent in the environmental movement. One of the groups in this field, the Natural Resources Defense Council, claims to go about achieving its goals in the following ways[19]:

Our primary strategies are:

- scientific research (determining the facts)
- public education (getting the word out)
- lobbying (persuading Congress to act)
- litigation (suing the bad guys!)

The first three of these strategies are of course "classic" for scientists' public-interest associations. The fourth strategy, however, signifies the insertion of legal expertise and personnel. The resulting comprehensive agenda thus draws on the contributions of both scientists and lawyers.

The NRDC and similar organizations (notably the Conservation Law Foundation, the Environmental Defense Fund, and the Environmental Law Institute) are hybrid voluntary organizations of lawyers and scientists that aim at strengthening their overall strategy by combining scientific expertise with legal know-how. Founded in 1967, EDF claims to have been "the first group to take scientific evidence into courts to achieve environmental goals" (source: EDF's Web site).

The profusion of environmental regulations has also triggered a small but vigorous countermovement of scientists' public-interest associations who prefer a less interventionist government, and criticize many measures as hasty and based on flawed science. For example, the Science and Environmental Policy Project questions the existence of a human-made global warming trend, and therefore opposes regulatory measures intended to alleviate such a trend.

In addition to working in those organizations with an express focus on a legal strategy, lawyers are increasingly useful to many other scientists' organizations, as legal options are becoming more pervasive. In this regard, the advent of specialized lawyers is another element in the professionalization of scientists' public-interest organizations—alongside the public-relations specialists, political consultants, and professional fund raisers. (Originally, no lawyers were invited to the Asilomar Conference of 1975; nonetheless, some lawyers were present, and they exposed the scientists to their very distinct perspectives on the topic at hand.)

Public-Interest Organizations and Scientific Organizations

Organizational sociologists know it, and veteran citizen-scientists know it, too: Creating functioning organizations is hard. Instead of building up a new organization from scratch, it is often much more efficient to get an existing organization to adapt one's goals. This makes existing organizations valuable resources for activists, or even targets for takeovers. Consistent with this general principle, some citizen-scientists indeed attempted to influence or widen the portfolio of concerns of existing scientific associations—or parts or special programs thereof—as well as science-related units of the federal bureaucracy, such as programs of the National Science Foundation.

Those existing scientific organizations include the disciplinary associations, such as the American Physical Society and the American Psychological Association, as well as organizations with a wider constituency, such as the American Association for the Advancement of Science. Citizen-scientists have tried to work within the professional organizations, but these often proved reluctant to take public stances on "non-scientific" issues. The debates about whether the scientists' disciplinary associations should enter the political arena (and thus effectively become, for some causes, public-interest associations) raised central questions about the boundaries between the scientific and the political system, and about the legitimate links between them. (We will take up these issues below.) The original reluctance of the scientific associations in some cases precipitated the formation of distinct scientists' public-interest organizations.

In the three following examples of major scientists' public-interest organizations, Moore (1996) documented how citizen-scientists first tried to influence these existing associations, and that only once the response was not as they had hoped for, they embarked on creating their own organizations.

In the late 1950s, Barry Commoner initially attempted to use the existing Social Aspects of Science Committee of the AAAS to formalize a program for disseminating scientific information to the public. When that committee balked at his proposals, Commoner and his colleagues at Washington University in 1958 formed the Greater St. Louis Committee for Nuclear Information. This group was the forerunner of the Scientists' Institute for Public Information, founded in 1963, which combined several local groups that disseminated information about fallout from the testing of nuclear weapons.

Similarly, Scientists and Engineers for Social and Political Action was founded only after the American Physical Society rejected the proposal that it should assume also a political role. In the fall of 1967, Charles Schwartz, of the University of California at Berkeley, proposed an amendment to the APS's constitution that would have allowed the members of the society to vote on any issue of concern. The concrete purpose of this amendment was to let the APS take a public position against the Vietnam War. The amendment was highly controversial, and the members defeated it by a margin of nearly three to one in June 1968. Consequently, Charles Schwartz, Martin Perl, Marc Ross, and Michael Goldhaber founded SESPA (originally Scientists for Social and Political Action) at the New York meeting of APS in February 1969.

An analogous pattern can be found in the relationship between citizen-scientists and their academic host institutions. Among the roots of the Union of Concerned Scientists were some MIT scientists' opposition to work on missile guidance being done at MIT's Draper Laboratory and the MIT administration's reluctance to take a public stand against the war in Vietnam (Leslie 1993; Nelkin 1972).

The relationship between scientists' disciplinary organizations and public-interest organizations is complex. As we have just seen, some scientists' public-interest organizations were founded after the citizen-scientists felt that the existing disciplinary organizations were not responsive enough to their concerns. Yet once the scientists' public-interest organizations had come into existence, they did create a ripple effect that extended back to the scientists' professional associations and even to the governmental science structure. In response to the scientific activists and their groups, professional associations, such as the AAAS and numerous disciplinary organizations, founded divisions or other subunits that addressed the social implications of science, albeit in a more moderate way. Once created, these sections of professional societies became some of the best-known, most effective, and most enduring vehicles for scientists' civic concerns, having had an institutional home from the start, and a certain continuity of support and funding that comes with it.

Let us examine what happened in the two mentioned organizations after they rebuffed the citizen-scientists' initiatives. In 1972, only 3 years after the activists had founded SESPA/SftP outside the American Physical Society, the Forum on Physics and Society was instituted within the APS. Its express purpose was to address issues at the interface of physics and society. Initially, the APS Council put the new Forum (then a new and unusual type of organizational subunit for APS) under special surveillance—probably an aftereffect of the provocative actions for which some members of Science for the People had become known. Yet the fears of some of the APS leadership proved unfounded, and the Forum on Physics and Society soon turned into a more "regular" part of the APS, to which about 10 percent of the total membership of the APS now subscribe.

The American Association for the Advancement of Science had actually had a long history of dealing with issues of the interrelationship between science and society.[20] The AAAS Committee on the Social Aspects of Science was founded in 1957. Although it did not accommodate Barry Commoner's proposals, in 1958 Commoner became chairman of the Committee on Science in the Promotion of Human Welfare, as the successor of the Committee on the Social Aspects of Science was called. Commoner thus worked simultaneously both on the inside and the outside of one of the established scientific associations. This was indeed the typical pattern, because the members of the scientists' public-interest associations would of course remain in their professional associations.

At least partly as a result of the efforts of citizen-scientists among its membership, the AAAS greatly strengthened its emphasis on the interface of science and society. An informal Science and Public Policy Study Group, initiated in 1966, became instrumental in helping academic programs in science and technology policy establish themselves. In 1970, the AAAS collaborated with NAS on a study (headed by Matthew Meselson) of the effects of defoliants in Vietnam. In 1975, the AAAS Committee on Scientific Freedom and Responsibility issued Scientific Freedom and Responsibility, a report prepared by John T. Edsall. A year later, the Standing Committee on Scientific Freedom and Responsibility was established. There was also, at the same time, a push within the AAAS to revise its constitution. In 1977, the constitution was amended to include among the objectives of the AAAS "to foster scientific freedom and responsibility, to improve the effectiveness of science in the promotion of human welfare."

Another major public-interest focus of the AAAS, and of other professional scientific organizations, has been concern for human rights in general, and particularly for the human rights and scientific freedom of colleagues who have faced oppression in foreign countries. To work on these issues, the AAAS has created a Science and Human Rights Program. Another institution active in this area is the New York Academy of Sciences; its Committee on Human Rights of Scientists was founded in 1978. The American Physical Society's Committee on International Freedom of Scientists exemplifies the similar efforts by numerous disciplinary associations. Professional societies have found it relatively easy to become active in the area of human rights because of its non-controversial nature. Nobody among the scientists or the wider American public would consider the activity of supporting these rights unworthy.

The ripples that emanated from the public-interest organizations as the centers of second-wave activism extended beyond the scientific associations, and even reached parts of the first ivory bridge. Institutions that have customarily been close to the government, such as the National Academy of Sciences and its related institutions, also vitalized or created organizational substructures dedicated to dealing with the social ramifications of science. Together with the National Academy of Engineering and the Institute of Medicine, the NAS has a Committee on Science, Engineering, and Public Policy (COSEPUP), which was founded in 1961 as the Committee on Government Relations.[21] COSEPUP became the vehicle of momentous investigations into the societal impacts of science and technology. Strategically placed in the vicinity of the government but independent of it, it has served as a crucial link between the first and the second bridges. The

COSEPUP studies that foreshadowed the Press-Carter Initiative were partly commissioned by Congress but partly undertaken on the initiative of its members and with funding from a private foundation. The NAS also has a Committee on Human Rights, founded in 1976, which since 1994 has been co-sponsored by the National Academy of Engineering and the Institute of Medicine.

The ripples even reached governmental science units. Price (1979: 86), for instance, mentioned a "Science for Citizens" program of the NSF, which supported public-interest groups to employ scientists for work on public policy issues (NSF 1976). As a result of deliberations that began in 1973, the NSF also instituted the "Ethical and Human Value Implications of Science and Technology" program, for which the first awards were given out in fiscal year 1976.

Relationships among Scientists' Public-Interest Organizations

In the economic jargon of the rational-choice theory of social movements, a group of social-movement organizations all of which follow the same overarching goal (e.g., protection of the environment, peace) is called a "social movement industry" (McCarthy and Zald 1977; McAdam, McCarthy and Zald 1988: 718). There are interesting questions about the relationships among the associations in such an "industry" (Barkan 1986; Morris 1984; Rupp and Taylor 1987). How strongly have the organizations (in our case, the scientists' public-interest associations) been connected with one another? To what extent have they collaborated? To what extent have they competed?

Among the scientists' organizations, there is a substantial overlap in membership. Consistent with the general sociological finding, mentioned above—joiners tend to keep joining—a sizable group of citizen-scientists participate in more than one association. The overlap exists not only at the level of leadership, but also at the level of support. Henry Kendall (personal communication, February 9, 1999) spoke of a fairly consolidated mailing list of about 3 million individuals in the United States who receive solicitations from nearly *all* the major scientists' organizations. Whereas scientists' organizations thus compete for funds from this pool, they do not compete, according to Kendall, in terms of contradictory or parallel goals and activities. Rather, in this respect, he perceived a division of labor between the organizations, each occupying a separate niche.

Successes and Limitations of Scientists' Public-Interest Organizations

A crucial yet somewhat neglected topic has to do with the outcomes of social movements. Following Gamson (1975), we should distinguish two kinds of outcomes. Outcomes of the first kind concern the status of the social-movement organizations—to what extent they have become recognized as legitimate representatives of their causes by the political system. Outcomes of the second kind concern their goals—to what extent they have been successful in promoting their causes. In regard to outcomes of the first kind, the scientists' associations have been doing well in general. Aided by the pervasive trends of internal professionalization and routinization, many scientists' associations of the third wave have become "players" in the Washington system of lobbyists and think tanks. A related indication of cooptation by the political system is that moderate goals prevail. The radical wing, never particularly strong, is at present even less visible. Yet we should not overlook the existence of a few groups that in some respects might be considered radical.[22]

To what extent have the scientists' public-interest organizations been successful? What did they achieve, and what were their successful strategies? Where did they fail, and why? Primack and von Hippel (1974) analyzed the conditions of success and failure for scientists' public-interest activities on the basis of six case studies. They distinguished between easy fights and hard fights. Fights were easy when no major vested interests were involved. In the hard cases, they identified the following crucial factors that determine success or failure: "the timeliness of the issue; whether it poses a personal and obvious danger to individual members of the middle or upper class public; the existence of an appropriate forum; the special visibility of certain issues in particular localities; and the credibility of the public interest scientists themselves" (Primack and von Hippel 1974: 240).

Successes

A thorough analysis of the degree of success of the voluntary public-interest organizations would necessitate case-by-case evaluations of individual associations and sufficiently narrow issues. It might nonetheless be possible, with reasonable plausibility, to discern some major impacts at the societal level. (We disregard the impact of activism on an individual scientist's life, although involvement in public-interest associations has, no doubt, produced great changes, including personal sacrifices, in the lives and careers of many citizen-scientists.) Here we intend to explore ways of thinking about this question, and thus to stimulate further research on this important issue.

The scientists' associations have, first of all, had crucial functions, and played a kind of elite role, within the ecology of the larger social movements to which they have belonged. Compared to the relatively small numbers of citizen-scientists, they have had a disproportionately great impact. One function has been to educate the movement and to provide the activists with pertinent information. Another function has been to generate prestige and credibility for the movements and to propagate their views and goals within the general public. The Emergency Committee of Atomic Scientists is one of the best examples for an organization that used its members' stellar scientific reputations to influence public opinion and to raise funds for the cause. The visible alignment of prominent scientists with the goals of a social movement transfers to that movement some of the respect the public holds for these individuals and for scientists in general.

The Union of Concerned Scientists is one of the success stories in terms of supporting larger movements. There is strong evidence that the UCS, which became involved early in opposing nuclear power, made an important contribution to the growth of this movement (Moore 1993: 257–259). The UCS experts, who highlighted the risks of nuclear power generation, were much more willing and able to present their views to the public than scientists with differing views were. As a result, the anti-nuclear faction among the scientists, which by many accounts was only a small part of the largely passive scientific community, gained a disproportionate influence in the public's perception.

In addition to the role the scientists' associations have played within the larger social movements, they also had an effect on other organizations and institutions located at various intersections of science and society, such as scientists' professional societies, academe, and science journalism. The existence of scientists' voluntary public-interest associations broadened the outlook of the traditional professional and disciplinary societies and even of some institutions of governmental science policy. As was documented above in more detail, the professional associations initially were very reluctant to enter the boundary regions of science and activism, but once autonomous public-interest associations were formed in response to that reluctance, the professional associations created their own organizational subunits to address public-interest issues, albeit typically in a toned-down fashion.

The citizen-scientists' activism also appears to have contributed to the establishment of "science studies" or similar programs in academic institutions. Apart from its intrinsic intellectual value, instituting such a program may have served, in some institutions, as a response to placate growing scientists' activism. Moreover, citizen-scientists were often available to create or staff such programs locally. For instance, citizen-scientists from MIT and other Boston area institutions were instrumental in founding the MIT Program in Science, Technology, and Society, a project greatly encouraged by MIT's provost, Walter Rosenblith, and its president, Jerome Wiesner, in the mid 1970s.

Furthermore, the scientists' associations collectively facilitated the growth of the field of science journalism (Nelkin 1987). This subspecialty of the media would probably have grown even if there had been no scientists' public-interest associations; yet scientists' organizations, which became quite sophisticated in providing relevant scientific information to the mass media, made science reporting easier and thus perhaps aided it in establishing its own critical mass. It certainly helped that the UCS was well set up to satisfy information requests from the growing community of science journalists—who to a large degree were themselves of the anti-nuclear persuasion (Rothman and Lichter 1982). Similarly, the Media Resource Service, which puts journalists in contact with scientists in a relevant area of expertise, has been a huge success. As mentioned, 30,000 scientists at present agree to be in the MRS's files. What can be observed for scientists' voluntary organizations has a more general parallel in the wider field of social movements. Communications have long been acknowledged as decisive in generating and sustaining collective action, and a certain interdependence has developed between social-movement organizations and the mass media (Gamson 1995). Movements try to use the mass media to get their message across to the public, whereas the media search for captivating stories to report.

Thus far, we have covered only effects on other social institutions; now the question of the efficacy of scientists' public-interest organizations needs to be tackled in a more direct way. To what extent have scientists' public-interest organizations furthered the goals to which they have dedicated themselves? An obvious indication that the activities of scientists' voluntary associations, of allied associations, and of individual scientists have not lacked beneficial impact is the fact that several of them have been awarded the Nobel Peace Prize.[23] Probably the most immediate gauge of success or failure is whether a voluntary public-interest organization that pursues legal strategies wins or loses lawsuits. In 1983, for instance, the Conservation Law Foundation initiated the legal action that resulted in the court-mandated cleanup of Boston Harbor. To give a more recent example of the same kind, the National Resources Defense Council settled a lawsuit it had brought against the EPA in early 2000; the settlement required the EPA to close some loopholes in the Clean Water Act. In such cases, public-interest associations are on safe ground when they claim success.

Sponsorship in elections also has an easily measurable outcome—the candidate wins or loses. This yardstick applies to organizations such as the Council for a Livable World that support political candidates. It is difficult to know for certain, of course, if the support by the CLW or similar organizations was decisive in the candidate's victory, or if he or she would have won anyway. But particularly if the race was close, such organizations will take pride in the election of "their" candidates who, in turn, will feel especially strongly indebted to their sponsors.

Many associations have focused their energies on supporting treaties or legislation. For the scientists' associations working in the area of peace and nuclear disarmament, international treaties have been crucial milestones. Some examples: the Nuclear Non-Proliferation Treaty (1970), the Anti-Ballistic Missile Treaty (1972), the SALT accord (1972), the Biological Weapons Convention (1975), the START I treaty (1994) and successive efforts), and the Chemical Weapons Convention (1997).[24] Pugwash, for instance, is commonly credited with helping to lay the groundwork for several of these treaties, including particularly the ABM Treaty, the SALT accord, and the Chemical Weapons Convention.

At the environmental front, laws or regulations take the place of treaties. To name but a few out of an impressive barrage of protective national legislation that has been put in place over the decades: the National Environmental Policy Act (1969), the Occupational Safety and Health Act (1970), the Clean Air Act (1970), the Federal Water Pollution Control Act Amendments (1972), the Endangered Species Act (1973), the Safe Drinking Water Act (1974), the Toxic Substances Control Act (1976), the Resource Conservation and Recovery Act (1976), the Clean Water Act (1977), the Comprehensive Environmental Response, Compensation, and Liability Act (1980), the Nuclear Waste Policy Act (1982), the Water Quality Act (1987), the Oil Pollution Act (1990), and the Pollution Prevention Act (1990). International accords to protect the environment (for instance, the 1997 Kyoto Protocol on combating human-made global warming, and forerunner efforts such as the 1992 "Earth Summit" in Rio de Janeiro) have also become increasingly important.

The establishment of the Environmental Protection Agency in 1970 can, at least in part, be considered a governmental reaction to demands from citizen-scientists. (By contrast, in the area of peace and disarmament, the governmental bridge—harnessing science for military purposes—was first, and the second bridge came into being largely in response to it.) Some of the voluntary associations have expended a great deal of energy in preparing draft legislation. In its early years, for example, the National Resources Defense Council focused on drafting major environmental legislation, such as the Clean Air Act and the Clean Water Act; and more recently, the Council for Responsible Genetics developed model genetic discrimination legislation that had an impact on political efforts in this area.

However, treaties and laws, such as the ones mentioned above, are not entirely reliable markers of effectiveness in an across-the-board way. Many factors influence their passage, and a close cause-and-effect relationship with the efforts of scientists' organizations often remains elusive. Case-by-case investigations of how particular laws and treaties have come about are necessary here. For instance, especially strong evidence for a scientists' organization having made a difference exists in the case of the Asilomar Conferences, which developed recommendations on rDNA policy that became the basis for National Institutes of Health guidelines.

Associations that try to influence the public or politicians through letter-writing campaigns and the like, or through public-relations initiatives, in many cases cannot be sure of how much they had to do with a favorable outcome. Organizations that focus on specific issues, particularly if these are new or formerly very obscure, make the most plausible claims of success. The Citizen's League Against the Sonic Boom is a good example because it was specifically directed against a concrete government plan—building an SST—and the plan was subsequently canceled, at least in part owing to CLASB's efforts. The Scientists for Sakharov, Orlov, and Shcharansky could feel gratified when the scientists they supported were released. The Center for Science in the Public Interest has been in the forefront of bringing previously little-known nutrition issues to the public's attention. When the Nutrition Labeling and Education Act of 1990 was passed, CSPI could plausibly claim a success. The Environmental Defense Fund can take a good share of the credit for publicizing the harmful impact of DDT on the environment and ultimately also on humans, and this campaign led to a nationwide permanent ban on DDT in 1972. By contrast, groups that work on pervasive and more general issues find it harder credibly to assert that they have made a difference.

Having reviewed the spectrum of activities on which scientists' associations may gauge success, we need to put these activities in perspective. They all are not actually ends in themselves; they are means toward the ultimate ends of security or a healthy environment, for instance. Hence, they do not automatically translate into "success," as defined in a more fundamental sense. The instituted measures may not always work as intended, or their effect may be dwarfed by stronger countervailing trends. For example, because the health of the global environment by some indicators is worse now than it was when the environmental movement started, the activists' influence—and the legislation they helped bring about—may have been only to keep the problem from being even worse than it has become. Detailed studies of the effectiveness and outcomes of the various activities are needed.

Arguably, the "bottom line" of the scientists' public-interest associations' success is the extent to which they have been able to shape public understanding, public attention, and the public agenda in a general way. For scientists' organizations, the greatest success may ultimately lie in influencing the public opinion about the role of science and scientists in society. By promoting their concerns to the forefront of public debate, the scientists' associations have played a crucial role in anchoring science-related issues of a general kind strongly in the public's mind (e.g., young people's awareness of the need for environmental responsibility).

This impact of course can add to, but potentially also detract from, the public's respect for science. One can imagine ways in which such a detraction might happen. For example, to a much stronger degree than the scientists' anti-nuclear-weapons activism, the scientists' environmental and anti-nuclear-power activism has been in a "strange bedfellows" relationship with anti-science and anti-technology currents in American culture that oppose nuclear power as part of a wholesale rejection of modern, technology-based life. The potential danger of such an association is that science might be perceived in some quarters as somehow abdicating its own claims to a scientific worldview.

Scientific knowledge in certain fields of public interest also might become regarded as so politicized that the public might turn cynical in the face of opposing truth claims and might no longer attribute the special quality of "objectivity" to scientific statements. If for every Linus Pauling criticizing nuclear testing, there is an Edward Teller defending it (Pauling and Teller 1958), the suspicion is likely to spread among the citizens that each side in a conflict simply recruits its own scientific operatives who support their patrons' interests through partisan evidence. Such a view would of course compromise a major traditional source for the public's respect for science and scientists. But none of these potential detractions has actually materialized to a large extent.

Limitations

Among the more concrete and historically observed limitations of scientists' associations' efficacy, Moore (1993: 169–170) noted an increasing fragmentation of the wider movement that prevented a united organizational super-structure from forming. There was never a unified institutional umbrella for the scientists' voluntary public-interest associations. Some degree of fragmentation is probably the corollary of a strong grassroots element of decentralized organizations.

Nichols (1974: 140–141) implied that it might have been more effective for the moderate scientists' public-interest groups to concentrate their efforts on the executive branch, where the largest part of the government-science interface is located, rather than primarily on the legislative branch. On the other hand, the executive branch in the early 1970s was so hostile toward citizen-scientists that the chances of success via that route might have been slim then. Another potentially relevant factor to consider is that, because the United States has no central Department of Science, federally sponsored science is supported, and science policies are made, in a number of different departments and agencies of the executive branch. This decentralized structure may have influenced the strategy or even the efficacy of scientists' organizations, yet it is unclear to what extent and in what ways.

The greatest drawback of scientists' organizations probably lies in the relatively small membership numbers within the scientific community. One of the failures of the scientists' voluntary public-interest associations, in Henry Kendall's (personal communication, February 9, 1999) estimation, is that they have been unable to break out of the "ghetto" of a distinct minority position within the scientific community. A major reason for this is that the social system of science has developed and now operates under a fairly autonomous normative regime that allocates resources mainly according to science-internal standards. As we have pointed out, if one wants to be a successful scientist (especially in academia) one had better dedicate oneself to doing good scientific research in an ivory-tower sense—"good" being defined by one's scientific peers' views of the intellectual merit of one's work. Achievements in other aspects are less important (perhaps with the exception, prominently at non-research-oriented colleges, of teaching). Specifically, activism in a public-interest association does not earn many bonus points, if any. This makes it less surprising that citizen-scientists have been in the minority, even at times when the waves of activism crested, compared to the larger groups of scientists who give scientific pursuits their undivided attention. The social system of science is set up in a way that internal criteria for success and allocation of resources prevail.

There are of course political, social, cultural, and environmental junctures that, at various times, boost or reduce the influence of the scientists' associations—which political party is in power, and with what sizable a majority; are there any well-publicized current crises; and so on. Another factor that, at first sight, limits scientists' influence is, again, the relatively small size of the scientific community as a whole within society. Scientists by themselves do not form a segment of the population that by its sheer numbers makes politicians listen, nor were scientists (including such prominent ones as Niels Bohr and J. Robert Oppenheimer) necessarily persuasive to politicians.[25] However, historically the scientists compensated well for their relatively low numbers. Among the main sources of the citizen-scientists' successes is that they possess crucial expert knowledge. The scientists' "lobby" is relatively weak when it asks for the public support of science; it is far more powerful when it commands expertise in questions that are considered vital by the public. The public's beliefs about the nature of this expert knowledge, as well as the beliefs of scientists, are crucial to the citizen-scientists' effectiveness.

Citizen-Scientists' Views of Their Ivory Bridge

What is the citizen-scientists' understanding of the ivory bridge they have been building? How do they define their own identity? In the following, we shall examine three distinct notions that are grounded in different views of science. We shall start with the predominant view—that science is objective, and should be used for the common good—and then also briefly visit the two minority views of science as "objective-partisan" and as "constructed."

Objective-Universal Science

Most moderate scientists' public-interest associations hold the view that the objective findings of science should benefit "the people" in general. Within this view, a core concern is how to define the borderline between objective science and activism for the common good. Fears of crossing this borderline have been a major factor in the scientists' professional organizations' reluctance to take an activist stance. Such an act, they feel, would compromise their commitment to scientific objectivity and introduce an element of political partisanship and strife.

This boundary issue, for example, dominated the heated debate among American Physical Society members about Charles Schwartz's constitutional amendment that would have allowed the APS to debate, and adopt a position on, any public issue. The commotion caused by the proposed amendment can be gauged from the numerous letters on the topic in the pages of *Physics Today* (a journal published by the American Institute of Physics) in early 1968. The opponents of the amendment—who eventually carried the day—tended to argue that the APS, as a disciplinary organization of physicists, should stay out of the political fray and should concentrate on matters of a scientific and objective nature. One contributor to this debate put it as follows: "To use our journals for very general discussion and debate means that they will lose their essentially professional character. Still further, the journals will fall into the hands of politically or socially oriented editors who will inevitably use them to support their own special viewpoints on matters far outside of the field of physics." (F. Seitz, *Physics Today*, January 1968, p. 17) Another letter warned: ". . . the raising of non-scientific issues—even if not adopted—could have a deeply divisive effect on the [APS] at a time when we need unity in the face of serious declines in relative numbers of students choosing physics for study and declining sources of support for basic research in physics. . . . Professional societies organized to run journals and meetings must not run the risk of alienating *any* professionally qualified member, for the societies have monopolies on the effective means of scientific communication." (L. Branscomb, *Physics Today*, February 1968, p. 13)

The boundary issue between science and politics rightly took center stage in these and other debates, because it has important societal repercussions for science (Gieryn 1983). As has long and convincingly been argued, the political influence and public respect scientists enjoy rest largely on the societal belief in the objectivity of science. In other words, the impact of scientists on society is at a maximum when the impact of society on science is believed to be at a minimum. The very trust in the independence of scientific truth from societal factors is the guarantor of the public's respect for the scientists' point of view—and thus at the very core of the scientists' associations' efficacy and success.

Most scientists' public-interest associations share with the professional associations and most members of the scientific community the core belief in the objectivity of science itself. The citizen-scientists try to steer a careful course that would allow them to contribute to the societal arena without giving up the notion of scientific objectivity. Moore (1996: 1593) noted that "activist scientists sought new ways to maintain credibility simultaneously as objective scientists and as political actors serving the public good." In the opinion of most citizen-scientists, their bridging project requires a solid ivory tower from which the authority of science can emanate to society. If that tower erodes, the ivory bridge collapses.

The public-interest organizations attack perceived "misuses" of science by industry and—at the first interface—by various government bodies. As an alternative, they advocate a socially responsible use of science and technology. They—often unconsciously—have operated in accordance with the classic liberal-democratic concept of the public or "the people" (Nichols 1974: 148). They never entered into a deep debate on the question of exactly who this collective subject, "the people," was. Typically the term has been used in a fuzzy universalist sense of "everybody." For the most part, the activists have worked to bring about benefits they considered ecumenical.

One of the main creeds that has helped moderate citizen-scientists negotiate the boundary of science and politics has been that the provision of pertinent scientific facts would enable the people to make better, that is, informed decisions. It is the civics textbook view of a democracy working through rational debates that are grounded in knowledge of the relevant facts. In particular, this was the line of the Scientists' Institute for Public Information, which consciously restricted itself to presenting expert scientific knowledge to the people so that they could decide for themselves in an informed way. Barry Commoner himself ran afoul of this restrictive mission of SIPI, and was ousted, when he declared his candidacy for the presidency of the United States in 1980; as a political candidate, he obviously needed more programmatic content than merely the concept of providing scientific information.

Other scientists' organizations do advocate policy positions, which of course makes their negotiation of the boundary issues between objective science and political activism even more complex, because these organizations can no longer position themselves neatly on one side, but now straddle the very boundary whose maintenance is the perceived prerequisite for the organizations' legitimacy and efficacy. Citizen-scientists tend not to spend too much time theorizing about "wearing-two-hats," but one can sometimes hear the argument that certain policy positions derive so self-evidently and commonsensically from scientific findings that these policies themselves acquire the mantle of objectivity. In other words, the underlying value choices are considered so non-controversial and universal (e.g., being against nuclear war, or against the destruction of our planet's ecology) that the linkage from objective scientific facts to policies almost appears as a technical rather than as a political one. This argument mirrors one made on the first bridge, whose occupants also often prefer to present questions as technical, rather than as political ones. In the frequent collisions between these two arguments, neither has escaped unscathed, and the broad trend has been toward re-defining issues as political that once were considered technical (Nelkin 1984a,b).

In sum, it is crucial to note that moderate scientists' associations typically agree with a basic position of their critics in the professional associations. Both sides operate on the premise that science is objective. Where they diverge is that the proponents of "ivory tower"-style scientific disengagement from society feel that the special nature of scientific objectivity and neutrality must be protected by strong boundaries, whereas the citizen-scientists hold that their objective scientific knowledge can—and should—

be used in the political sphere to bring about societal benefits and to prevent societal harm. Again, in that bridging effort, the public's belief in the objectivity of science is a major source of the citizen-scientists' efficacy and success.

Objective-Partisan Science

The radical scientists' associations (notably Science for the People) also regarded scientific knowledge as objective; but they vehemently rejected the notion that science could be "neutral." They viewed science primarily as a tool used by social groups to increase their power. Therefore, these associations set out to castigate the misuse of science by the ruling class and to wrest science from its control. Rather than engaging in a critique of the epistemology of science itself, they aimed at putting science and technology in the hands of "the people." Their definition of "the people," however, differed from the understanding noted above, and took on a much more partisan flavor.

Organizational sociologists have found that sometimes those who are not oppressed themselves actively engage in working for the oppressed.[26] Some of the more radical scientists' organizations could be considered part of this category, as they aligned themselves with the poor, the ethnic minorities, and the Third World (which, in their view, was victimized by US imperialism). The theoretical underpinnings for this stance were more elaborate, often Marxism-inspired sociological theories that identified "the people" with "the working class" and allied segments of the population. These radicals regarded the perceived misuses of science not as isolated aberrations within an overall acceptable status quo, but as a systemic problem in late-capitalist society, to which they were deeply hostile. Within this context, they also criticized the hierarchies, reward systems, and discriminations that existed in the social organization of science.

Aside from their sociological theorizing, the large amount of which was uncommon for scientists' associations, members of Science for the People were also very interested in activities of a practical nature. Through its Technical Assistance Program, Science for the People provided support to communities and political groups who fell in its re-defined category of "the people," to help them advance in their class struggle. A 1970 issue of *Science for the People*, for instance, reported that SftP activists installed an electrical generator at a free community medical clinic that lacked electricity. Moreover, the article went on, they helped the Black Panther Party "in evaluating, purchasing, and maintaining a truck and in setting up outdoor sound and communications equipment. . . . In every instance technical helpers explain to the people receiving what they are doing and why." (cited in Moore 1996: 1619)

Socially Constructed Science

A group of female members of Science for the People brought about a crucial development within that organization around 1970. The women started to examine the nature of science itself, not just its societal uses and abuses, and they claimed that the existing science was inherently and epistemologically biased. In 1974, SftP women edited a special issue of *Science for the People* that probed various avenues to a feminist transformation of science.

Although this strand of thinking remained a minority position among the SftP activists, it continued to find an outlet in the pages of *Science for the People*. Leanna Standish (1979), a physiological psychologist, advocated the formation of women's science collectives. In "A Feminist Critique of Scientific Objectivity," Elizabeth Fee (1982: 31) challenged basic concepts of traditional science. "If we are to move in the direction of a more fully human understanding of science, we should resist rigid separations between the production and uses of knowledge, subject and object, thinking and feeling, expert and non-expert. This requires re-admitting the human subject into the production of scientific knowledge, accepting science as a historically determined human activity, and not as an abstract autonomous force." Ruth Hubbard (1986: 20) argued in a similar vein: "What feminists have to contribute [to natural science] is the insistence that subjectivity and context *cannot* be stripped away. They must be acknowledged if we want to understand nature and use the knowledge we gain without abusing nature. Natural scientists must try to understand our position in nature and in society as subjects as well as objects."

These women blended feminist ideas into the SftP beliefs, denouncing what they regarded as a deep-seated male bias *within* the methods and epistemology of science, and proposing a different kind of science. In their reconceptualization of science, this group thus brought about, within a small segment of the scientific community, a synthesis of radical feminism and social constructivism—two intellectual currents that had been developing more noticeably outside of the science community.

The most far-reaching critique of science itself was taking shape in many of the academic programs of Science, Technology, and Society, or Science Studies, that sprung up at universities during the second wave of scientists' activism. Many of the scholars in these programs made the social construction of scientific knowledge itself their central topic and became critical of the notion of "objective scientific facts."

Especially with the demise of Science for the People, the women's groups among the scientists' voluntary organizations are usually not concerned with the establishment of an alternative science that conforms to "women's ways of knowing" (Belenky et al. 1986). Rather than subscribe to the radical feminist critique of science, these groups accept the general principles and methods of science as currently known, and concentrate on eliminating overt and subtle gender discrimination and on improving women's participation and prospects in science careers.[27]

The Twilight Of Cincinnatus?

The "politicization of science" that occurred in the course of the two linkage projects also supported the growth of a new class of experts *on* science—science policy managers and analysts, science journalists, public-relations and fundraising specialists, science lawyers, and so on. These experts introduced canons of knowledge, skills, and sophistication with which the association-forming scientists were usually unfamiliar. The founders of the scientists' public-interest associations had acted as citizens in Cincinnatus-like fashion, but the developmental trends in scientists' voluntary public-interest associations have moved them away from that ideal.

There now exists a stable second bridge of scientists' associations. Along with their obvious benefits, the processes of institutionalization also have a potential downside for scientists' groups. They may lose something of their unique appeal—the air of Cincinnatus or of the film *Mr. Smith Goes to Washington*, which certainly still strikes a cord in the American psyche. They are in danger of coming closer to being identified, in the eyes of Washington insiders and of the general public, as just another pressure group. The difference between the two interfaces of science and society has certainly diminished, owing to the widespread processes of bureaucratization and professionalization that make the second one look increasingly similar to the first.

The professionalization of scientists' activism might also hold certain dangers for the independence and authenticity of the citizen-scientists. To the degree that they make activism a career and become "established dissenters" in a public-interest association, they might be seen to turn into soldiers of their cause and into mirror images of their counterparts, the scientist-administrators.

The history of the scientists' organizations thus plays out the dilemma of the citizen in the age of professionalization and specialization. Amateurish citizen-scientists formed these organizations with the express purpose of bridging the gap between their ivory tower of specialized knowledge and a society which is affected by it, but their scientific expertise came to be complemented by many other kinds of professional expertise. Such professionalization may have been necessary in view of the massive professionalization of the governmental interface between science and society, and it certainly has been beneficial (see Salomon 2000). It has helped many scientists' public-interest associations to improve their efficiency and capability in advocating their causes. This might make successes and victories more likely, to the good of humankind.

Although many scientists' organizations have become more professional, this trend has not totally transformed the character of the organizations. There is still ample room, and a great need, for citizen-scientists acting in the spirit of volunteerism to make valuable contributions, not only at the rank-and-file level but also in leadership positions. Moreover, the history of scientists' voluntary public-interest associations has shown that existing organizations are often crucial preconditions and valuable assets for an emerging social movement or wave of activism. With a multitude of scientists' associations already in place, new issues at the interface of science and society will have good chances to find the organizational support that is crucial in publicizing and promoting them.

4

Autonomy and Responsibility

At the end of our survey of the two political bridges that link science with society, we return to the theoretical framework outlined in the introductory chapter: the systems theory's view of modern society. This perspective permits us to understand the seemingly disparate interfaces as homologous and complementary phenomena. The two bridges certainly differ from each other, they often stand in opposition to each other; but we realize that the two bridges—topics that are not usually treated in conjunction—are in fact similar: They both are attempts to coordinate a highly autonomous science system with society through political pathways. What we have called bridges are, technically speaking, media of exchange between social systems. Mastering this challenge of coordination, without de-differentiating society again, promises opportunities for advances in multiple areas of life that are impossible in societies of a simpler structure. Here, we have contended, lies the explanation of the paradox we discussed at the beginning of the book. The existence of an autonomous science system has enabled science to progress faster and more decisively than it would have under tighter external controls. Parts of the enormous scientific potential thus created have then been transferred into society via the various exchange media. In that way, an autonomous science system has allowed both science to flourish and society to reap benefits. A promising strategy to generate even greater societal benefits is to improve the linkage mechanisms (no trivial task).

A common way in which modernizing societies have addressed the problems of increasing differentiation, and of keeping the more numerous and more complex subsystems together, has been to have the government coordinate the subsystems, and, in extreme cases, to have government dominate all aspects of life. The mushrooming of government functions is often fraught with challenges. According to an interesting sociological analysis of contemporary society, societal crises may be caused when government intrudes into previously private areas of life—previously private areas that, one should note here, had only been created in the course of societal differentiation (Brand 1982; Habermas 1975, 1981; Kreckel et al. 1986; Sonnert 1987).

In this view, the crises triggered by government expansion often precipitate a particular class of social movements—termed "new social movements." Not only do these movements react to the process of politicization of private life, brought about by government intervention; but, through their own reaction, they also propel this very process, because they often advocate an even more thorough politicization (though perhaps of a different kind). For example, once the government enters the field of health care through direct assistance programs or regulations, health care turns from a private into a more political issue. Now groups and constituencies who feel disadvantaged or unfairly treated by the government are likely to enter the political arena to voice their demands, and this tends to politicize health care even further.

According to this view, new societal fault lines, and new social movements come to the fore, while expansion of the welfare state, in concert with rising levels of affluence among large portions of the population, weakens the once-dominant economically based class conflict between the poor and the rich, or between the "working class" and the "bourgeoisie." The varied group of the new social movements attracts mainly two types of members, those who are oppressed and marginalized by the existing welfare state society, and those who oppose the status quo, even though they possess a comfortable status within society (Sonnert 1987).

When we apply this general theory to the present context, the "politicization of private life" appears as a "politicization of science." The governmental involvement in science, which has dramatically increased since the start of World War II, would correspond to the other expanding government functions; and the scientists' associations would correspond to the new social movements, as they react—in different ways—to the strengthening nexus of governmental politics and science. Because of their position within society, the citizen-scientists would clearly belong to the "new middle-class activist" type of new social movements.

In addition to the growth of the interventionist state, a related development may further advance the politicization of science and scientists. It has been argued that, in modern societies, political conflicts are increasingly about the distribution of risk rather than about the distribution of wealth (Beck 1992). In such a scenario, expert knowledge about risk—of the kind scientists might possess—becomes a pivotal political resource, and scientists thus are inevitably drawn into political disputes.

These theoretical considerations make clear that the two paths from science to society are strongly interrelated. We now briefly recapitulate the history of their interrelationship. When, in the mid 1940s, both media of exchange emerged in a major way, they were not yet clearly differentiated. Many first-wave citizen-scientists were also closely involved in the emerging field of governmental science policy, and neither pathway was yet highly professionalized. During the second wave, the two bridges were separated from each other by the widest distance. Numerous second-wave citizen-scientists saw themselves as an oppositional force, emphatically refusing to become part of the "system," and even scorning scientist-administrators as corrupted sellouts. Whereas governmental science policy had become more professionalized by that time, many citizen-scientists ardently repudiated going down that road.

More recently, however, scientists' activism has been catching up with the professionalization trend in science policy. On the one hand, many disciplinary science associations, with coordination provided by the American Association for the Advancement of Science, now offer fellowships for scientists to work in Washington, get acquainted with the process of making science policy, and glimpse into the world of scientist-administrators; on the other hand, many scientists' public-interest organizations have turned into lobbying groups. In extreme cases, citizen-scientists have become quite similar to scientist-administrators. Many individuals are now willing to be active, consecutively or simultaneously, in both governmental science policy and scientists' public-interest activism. In the environmental domain, impulses from scientists' activism have led to the institutionalization of a rich regulatory and bureaucratic structure that addresses the concerns of citizen-scientists and environmentalists. In sum, we observe a re-convergence of the scientist-administrators and the citizen-scientists under the auspices of increasing institutionalization and professionalization. The two exchange media between science and society have now become so integrated that one might even perceive the outlines of one larger exchange mechanism that comprises both original bridges.

So much of past developments; but both bridges are still under construction and constantly evolving. This prompts the question: What lies ahead for them? As to the first bridge, in the post-Cold War era, the traditional ways of governmental support for science and technology may need updating. The strong and persistent antipathy within the American political culture against supporting an establishment that is not accountable, militates against anything that could be misconstrued as a "blank check" for scientists who are seen by some as indulging in their own esoteric interests with taxpayers' money. An explicit re-conceptualization of federal science policy may alleviate this latent threat to the governmental support of scientific research programs whose uses are not widely and clearly understood. The traditional notion of a dichotomy between Newtonian science ("curiosity-driven" basic research undertaken without regard to societal benefits) and Baconian science ("mission-oriented" applied research that harnesses known science to solve practical problems) is unfortunate and perhaps even dysfunctional for a flourishing science program supported by the federal government.

In this book we have proposed, within a comprehensive federal science policy that continues to support adequately both basic and applied research, the explicit and conscious additional funding for what we call Jeffersonian Science, that is, basic scientific research in areas of defined societal need. A governmental program of Jeffersonian Science would tie basic science tightly and comprehensibly (for the public and policy makers) to the national interest, and thus may have a good chance to invigorate and strengthen federal support for science, while supporting the autonomy of scientific research. Increased scientific attention to the current problems of modern life may be our best hope of solving them. Science is ready to do much more for society than it is currently being asked to do. There is a potentially rich and rewarding Jeffersonian agenda. Public concerns about creating a new establishment of scientists are best met by showing that scientists can be even more beneficial to society, and that their work can be most beneficial if it is not kept on too short a leash of expected social relevance (Branscomb 1999; Branscomb, Holton, and Sonnert 2001; Holton 1998; Holton and Sonnert 1999).

As to the second bridge, we have noted that, in parallel with the pervasive trends of institutionalization and professionalization, many of the scientists' voluntary public-interest associations have become more moderate, and are now, typically, members of the more comprehensive structure of science policy making that encompasses them as well as the governmental infrastructure. This tendency, however, need not be immutable in the future. Some of the scientists' public-interest organizations might easily turn again more "radical." The new Internet-driven communication patterns might even make the situation more volatile, because computers with Internet access can help make obsolete a great deal of the conventional bureaucratic and logistic structures that scientists' public-interest associations needed in the past. It is still unclear if the violent outburst of citizens' activism at the 1999 Seattle conference of the World Trade Organization was an isolated incident or a harbinger of things to come. Scientists' associations have not participated in this perhaps emerging trend of more radical activity; but if that trend indeed became a major wave, it might also encompass parts of a newly re-radicalized wing of the scientific community and citizen-scientists.

At the level of issues, however, it is already clear that a few citizen-scientists' groups are poised to participate in a wider anti-globalization movement. For example, the Council for Responsible Genetics and similar groups have begun to contest genetically modified crops and the biotechnology industry, where global capitalism and scientific research are most provocatively fused.

If there is a new wave of grassroots activism—akin to the crest of the second wave in the late 1960s—citizen-scientists will be able to take advantage of an already existing organizational infrastructure, which may greatly enhance their efficacy. As pointed out throughout the book, existing organizations are a most valuable resource for citizens' activism. Thus the now existing collective of scientists' public-interest associations would bode well for citizen-scientists who wish to make their voice heard more aggressively in the public arena. Perhaps ironically, the anti-globalization movement, if it is to gain force, will draw one of its main strengths from the globalization of modern communication technology.

One of the important questions that our study raises is that of the role of the scientific elite versus the citizenry at large. This is an instance of the ubiquitous tension between expert knowledge and democratic decision making in modern society. Whereas the chief function of scientist-administrators has been widely and fairly unambiguously understood as supplying scientific expertise to officials in the federal government, the citizen-scientists' thinking about this issue has followed two main models:

the discursive model of decision making in which the citizen-scientists as experts merely provide relevant information as input to the democratic process[1] and the more problematic commonsense-surrogate model of decision making in which the issues are deemed so commonsensical that universal democratic agreement can be theoretically derived or anticipated. This anticipation then legitimizes pressure-group activities, in which scientists' associations try to influence the political process.

Another deep issue is the dichotomy of isolation and connection—autonomy and responsibility. It has been a theme of this book, symbolized by the metaphors of the ivory tower and the ivory bridges. This fundamental topic is addressed at both interfaces, but the emphasis is characteristically different at either one. The priority of science policy has clearly been to harness science in the national interest. Social usefulness is the ultimate justification for spending federal funds on science support. The stance of many scientists' public-interest organizations toward the autonomy of science has been ambiguous. On the one hand, there is the wish to protect the independence of scientific research from governmental, especially military, influence—or exploitation, as some would probably put it. This viewpoint was strongest at the crest of the second wave, when the Vietnam War put a particular strain on the relationship between university scientists and the military research sponsors. In fact, one of the pivotal events—the March 4, 1969, work stoppage at MIT—can be seen as the symbolic expression of this rift. Many citizen-scientists abhorred the way the government used science and science-based technologies. On the other hand, the scientists' organizations did not just call for severing the ties between science and society, and somehow resurrecting an independent ivory tower of science. A return to scientific isolationism along the lines of G. H. Hardy (see chapter 1 above) was never a seriously considered option. Instead, they have demanded a better and more responsible use of scientific knowledge by the citizenry as well as the government. Both these stances also distinguished the scientists' groups from the anti-science tendencies in the larger countercultural movement of the time. Adherents of anti-science beliefs considered science itself—perhaps even rationality itself—as a force for evil, and wanted to blow up the ivory tower along with its bridges.

At a personal level, many scientist-administrators tend to focus on protecting their autonomy as scientists in the face of the job demands of being a loyal team player within the government. This kind of autonomy has traditionally been less of a concern for citizen-scientists. While the scientist-administrators are typically being hired and paid for their work at the intersection of government and science, the citizen-scientists are still often volunteers who could walk away from activism more easily. Many choose a path on which sacrifice is more certain than remuneration—although, as we have observed, scientists' activism itself is becoming more of a career. There may even come the day when some citizen-scientists will feel that they, too, have to protect their scientific autonomy in the face of loyalty demands from a social movement, or in other words, to resist turning from activist scientists into activists.

The personal difficulties and potential role conflicts in defining an appropriate code for scientists' behavior at the interface of science and society, are related to the following, broader questions: How does a person, at the individual level, deal with the fact that societal modernization intensifies the division of labor and brings about concomitant processes of compartmentalization and fragmentation? How are these societal trends related to psychological developments?

The division of labor in modern society enables many scientists to do science autonomously, following science-internal directives—based on the truth imperative—in their professional behavior. In an autonomous science system, societal consequences of this truth seeking appear as largely extraneous considerations. However, a countervailing force within some individuals resists such compartmentalization. Although modern complex society certainly presents unique challenges to identity formation, some people are still able and willing, particularly at high levels of moral and ego development, to look beyond the boundaries defined by particular societal subsystems, and to synthesize their different and sometimes conflicting role expectations into an overarching sense of personal agency, as theories of adult development in various domains have argued (e.g., Kegan 1982; Kohlberg 1984; Loevinger 1976).

Popular culture contains the stereotype of the scientist as an undersocialized and awkward individual who is helpless in the face of "real world" problems. Whereas it is taken for granted that "nerd" scientists function at the highest cognitive levels in their research, they are not trusted to perform well when it comes to tasks outside their narrow area of expertise or to common sense tasks. This part of the "nerd" image, however, may be far from the truth, a few highly visible cases in the scientific community notwithstanding. For instance, a study of scientists' concept of the good life found that scientists were able to reason at the highest levels of complexity also in non-scientific domains (Lam 1994). Activist scientists would provide another example of scientists' thoughtful engagement with issues that transcend the scientific subsystem. Whereas ivory-tower scientists may claim reasonably that it is not their personal responsibility to find a solution to the dilemma of reconciling the truth imperative of science (to which they pledge their primary allegiance) with the societal repercussions of science and technology, citizen-scientists do not cede personal responsibility for the societal effects of science. They strive to replicate, at the individual level, a synthesis that, at the societal level, is formed through the interplay of the various subsystems. This attitude of responsibility, collectively bundled, has been a main factor in the scientists' civic activism.

From a crude functionalist point of view, one might speculate that society would run more smoothly if nobody had a "conscience" and everybody conformed to the demands of authority. Objections to this rather totalitarian scenario can be leveled from many directions, but the point to be made here is that even judged from its own functionalist perspective, this scenario is unsatisfactory. On the whole, a certain intractability of individuals and the resulting conflict between governmental science policy and scientists' activism, although it may look like a burden and distraction, may ensure that, in the interplay of opposing forces, a more functional path of development is eventually chosen, because the process, when it works, appears to avoid the potential myopia of a single, isolated decision-making body.

We are convinced that the autonomy of science is necessary for its continued health and progress—and for its most valuable contributions to society. Supporting only research with immediate "social relevance"—in effect razing the ivory tower—would be short-sighted and counterproductive, not only for science but also for society. We have argued that to better protect the autonomy of the ivory tower, one must connect it with society through better ivory bridges. Disregard for the potential societal benefits and pitfalls of scientific progress would threaten to undermine public and governmental support for science, in addition to causing avoidable risks, dangers, or missed opportunities.

Of course, there must still be full support and regard for those scientists who devote themselves exclusively to basic scientific research. Not all scientists would wish to, could, or indeed should, be actively involved at the intersection of science and society. The Vietnam-era document calling for the 1969 MIT work stoppage observed that "the concerned majority [at that time] has been on the sidelines and ineffective" (cited in Gottfried 1999: 44). Perhaps this is as it must be. But we hope our attempt through this project to understand better how scientists' voluntary public-interest organizations came about and acted, succeeded or failed, is not only a historical and scholarly contribution, but may also help to open more minds to that Tocquevillian phenomenon at work in the sciences: taking the opportunity to join with one another, in order to fulfill one's wider civic responsibilities. It recognizes the efforts, vital to both science and society, that have been undertaken by scientists without monetary rewards, without helping their careers, and mostly laboring in relative obscurity.

One might disagree with specific goals, proposals, and demands of specific scientists' voluntary organizations. The following appendix of scientists' voluntary public-interest associations and related organizations (appendix D) will amply demonstrate that they are not all alike and do not all agree on the same program. Yet generally speaking, they have been doing an important service to our society—one that has not been sufficiently noted and appreciated by society at large. They have prevented science from becoming either too subservient to the demands of government or, at the other extreme, a new establishment in itself, and have preserved the image and the reality of the scientist as beneficent dissenter.

Appendix A
A Concise History of the Presidential Science Advisory Structure

To provide some background information for the discussion of the "first ivory bridge" in chapter 2, this appendix briefly sketches the institutional development of science advising to the president. For more details, see Barfield 1981; Beckler 1976; Branscomb 1993; Bromley 1994; Brooks 1964; Dupree 1986; Golden 1993; Herken 1992; Killian 1977; Kistiakowsky 1976; Smith 1992; Thompson 1986–1994; Trenn 1983.

William Golden's 1950 report to the president, titled Mobilizing Science for War, played an important role in establishing a science advisory structure in the White House. (For the entire text of Golden's report, see Blanpied 1995: 65–67.) Yet Golden's proposal of a science adviser and a small advisory committee encountered powerful opposition, especially against the position of a Presidential Science Adviser. Golden's plan was watered down in the process. In April 1951, a ten-member Science Advisory Committee was established within the Office of Defense Mobilization; it did not directly advise the president. That committee's chairman (the first was Oliver Buckley, former president of Bell Laboratories) also did not report directly to the president (Herken 1992: 55–57), although de jure he could have acted as the President's Science Adviser according to the plan Truman had approved (Trenn 1983: 38). (For the text of Buckley's appointment letter, see Blanpied 1995: 724.) Buckley was succeeded in the Science Advisory Committee chair by Lee DuBridge and I. I. Rabi.

In 1957, Sputnik shocked the nation and swiftly led to a substantial upgrading of the science advisory structure. President Eisenhower created the Office of Special Assistant to the President for Science and Technology (a position commonly known as the "science adviser") and appointed James R. Killian to it. Killian was succeeded by George Kistiakowsky (Beckler 1976). Eisenhower also reconstituted and invigorated the old Science Advisory Committee as the President's Science Advisory Committee (PSAC) (Wiesner 1993). Among the first members of that committee were Robert Bacher, William O. Baker, David Beckler, Lloyd Berkner, Hans Bethe, Detlev Bronk, General James Doolittle, James Fisk, Caryl Haskins, James Killian, George Kistiakowsky, Edwin Land, Edward Purcell, I. I. Rabi, H. P. Robertson, Herbert York, and Jerrold Zacharias (Herken 1992: 106, 253). Thus, about 7 years after Golden's initial report, his vision had been realized.

During the Kennedy administration, a decisive step was taken toward further institutionalizing the science advisory structure. President Kennedy's Science Adviser was Jerome Wiesner of MIT, whose unusually close and personal relationship with the president was due to their having become acquainted earlier through Massachusetts politics. Kennedy's 1962 Presidential Reorganization Plan, which was based on Richard Neustadt's study of White House Organization, created the Office of Science and Technology (OST). The reorganization authority under which Kennedy established the OST made it a part of the Executive Office of the President. The OST had to get its own line item in the budget, and its director was subject to Senate confirmation (Brooks 1964). It thus was under a measure of Congressional scrutiny and control. The Science Adviser, who was now also the Director of the OST, in that respect lost the executive privilege that applied to a Presidential Assistant and was required to testify before Congress. Through the reorganization, the Science Adviser and the OST staff acquired the added function of coordinating the sprawling federal science and technology bureaucracy. To this day it is an issue of debate which institutional arrangement—being only a Special Assistant to the President, or (simultaneously) head of a more regular government unit—would allow the Science Adviser to have a stronger impact on policy. Frank Press (1993), for instance, favors the first alternative.

Wiesner was followed by Donald Hornig, appointed by Kennedy only a week before the president's assassination. Hornig continued to serve in the Johnson administration. President Nixon's first Science Adviser was Lee A. DuBridge, who was followed by Edward E. David. Under Nixon the presidential science advisory system reached its nadir. The science community was increasingly split on fundamental issues, such as national defense and the Vietnam War, and voices critical of the administration also entered the science advisory structure. The PSAC advised the administration against its ABM buildup, and one member publicly opposed the administration's SST program. As a result, the relationship between the administration and its scientific advisers deteriorated. This alienation probably played a role in Nixon's decision in early 1973 to abolish the PSAC and the OST and to assign some of the functions of these bodies to the National Science Foundation. NSF director H. Guyford Stever officially became the science adviser, but the post had been severely downgraded in visibility, access, and authority (Trenn 1983: 86-93).

In 1976, during the Ford presidency, a modified science advisory structure was reestablished, this time by Congressional legislation (the National Science and Technology Policy Organization and Priorities Act). Ford had requested a legislative basis to preclude a future Nixon-style unilateral abolition. A successor the the OST called the Office of Science and Technology Policy (OSTP) was created, and the Science Adviser, who was restored to the old format, became also the director of that office. The institutional setting of the OSTP was again the Executive Office of the President, not the White House Office. In this reestablished institutional structure, H. Guyford Stever served as Science Adviser and Director of the OSTP during the waning months of the Ford administration. The legislation also provided for a President's Committee on Science and Technology (PCST) that, within two years, was to conduct a comprehensive survey of the federal science and technology effort. According to Press (1993), President Carter disliked the general idea of advisory committees, and this attitude also extended to the PCST. Apparently because the president did not want to keep a body similar to the PSAC around, his Executive Order 12039 abolished the PCST in 1977. The OSTP and the OMB were to take up its functions. To compensate for the lack of a permanent science advisory committee, Carter gave Press permission to convene ad hoc scientific committees.

Under President Reagan, George A. Keyworth II (until 1985) and then William R. Graham served as science advisers. D. Allan Bromley advised President George H. W. Bush, who upgraded the science adviser's office from "Special Assistant" to "Assistant to the President"—a significant gain in rank and prestige within the governmental hierarchy. In the Clinton administration, John H. Gibbons held the office until 1998, when Neal Lane succeeded him. In June 2001, President George W. Bush announced his intention to nominate John H. Marburger III as OSTP director and to make him his science adviser.

In 1982, Science Adviser Jay Keyworth set up a White House Science Council, which reported to him rather than to the president. In 1990, reviving the tradition of presidential science advisory committees, President Bush created the President's Council of Advisors on Science and Technology. Using a very similar name, President Clinton established the President's Committee of Advisors on Science and Technology in 1993. At the same time, Clinton established the National Science and Technology Council (NSTC), replacing the Federal Coordinating Committee on Science, Engineering and Technology. NSTC was a cabinet-level council, chaired by the president, that coordinated science policies across the agencies and drew on PCAST for advice, especially for advice from the private sector. In March 2001, President George W. Bush announced the formation of the President's Council of Advisors on Science and Technology.

Appendix B
List of Research Questions Assembled by Frank Press

The following list was attached to a memorandum from Frank Press and Bowman Cutter to the president and dated November 11, 1977. Titled "Examples of Important Research Questions of National Interest," it was labeled "TAB A." This list constituted the nucleus of the much longer "master list" of research questions that was the main product of the Press-Carter Initiative (appendix C below).

Can simple chemical reactions be discovered that will generate visible radiation? The results of research on this question may lead to inexpensive lasers for communication and industrial uses.

How does the material pervading the universe collect to form complex organic molecules, starts, and galaxies? Research in this area can provide increased understanding of fundamental natural laws and the origins of the universe.

What are the physical processes that govern climate? Greater understanding of climate could aid in the prediction of climate changes and allow time for measures to offset their impact.

To what extent is the stratospheric ozone affected by contamination of long-lived, man-made chemicals? The results of this research are important to man's survival and to the future of major industries.

What is the petroleum potential of the continental slopes and the adjacent ocean floors beneath deeper waters? This work is helping to identify the resource potential of the ocean's floor beyond the OCS.

How do organisms in the deep sea influence the productivity of the ocean? How will they react to sea floor dumping and mining activities? Answers to these questions will aid in assessing the future of the ocean as an important food source and should also provide baseline data on contamination of the sea.

To what degree can biological nitrogen fixation be enhanced? Successful research directed toward this question may provide more information on joint plant-bacteria relationships and an environmentally sound method of increasing crop productivity while minimizing energy costs.

What are the individual and cumulative effects of government regulation on domestic productivity? This research will provide a sound technical basis for assessing the benefits and cost of proposed, as well as existing, government regulations.

Can new homogeneous catalysts be prepared that will catalyze chemical processes important to the chemical industry? Research in this area could make it possible to make specific molecules needed in industrial processing techniques with minimum energy expenditure and without the creation of unwanted molecules that may pollute the environment.

What are the limits for communications use of the channel capacity in the visible spectrum? Progress in this area could significantly expand the capacity of optical communication systems, and since these systems use glass fibers instead of copper, their use would result in tremendous monetary and resource savings.

How do cracks initiate and propagate in materials? This research should provide information needed to develop structural materials that resist corrosion and failure under stress.

How do cells change during growth and development? Advances and understanding in this area should provide insights into the development of cell specialization and, perhaps, the aging process.

How do enzymes work? This research should help discover how enzymes selectively catalyze and control the chemical reactions carried out by living systems. The results of this research should extend knowledge on how to synthesize molecules in living cells.

What are the molecular mechanisms by which genes are regulated to produce specialized products, and what new information is required to exploit the new DNA recombinant technology? This work may lead to improved knowledge of gene action.

What are the factors controlling cognitive development? For example, how can the large number of component processes involved in reading and understanding a paragraph be characterized? Research on this question should provide new knowledge on the processes involved in reading and comprehending text. Such work is important on providing a basis for improving the techniques for teaching people to read and comprehend.

What are the mechanisms responsible for sensory signal processing, neural membrane phenomena, and distinct chemical operations of nerve junctions? Research in these areas will extend knowledge of perception, behavior, and the chemical functioning of the nervous system.

How can structures be designed and constructed to be both economical and earthquake resistant? In addition to reductions in life loss and personal injury, implementation of improved design procedures is expected to reduce losses to buildings alone by an average of $250 million per year.

Appendix C
Master List of Research Questions (OSTP News Release)

This list was the main product of the Press-Carter Initiative. The bracketed annotation at the end of each item indicates its origin—"Press" indicates that the item was in Frank Press's original memorandum; "DA" indicates that it came from the Department of Agriculture, "DoD" that it came from the Department of Defense, "DoE" that it came from the Department of Energy, "S" that it came from the Department of State, and "NASA" that it came from the National Aeronautics and Space Administration. The four questions submitted by the Department of Transportation but not included in the original list appear here under the heading "Engineering, Computer and Material Sciences."

OSTP News Release
Contact: Stanley D. Schneider

To: ALL SCIENCE WRITERS

In his public lecture to the American Association for the Advancement of Science on February 13, 1978, Dr. Frank Press, Director of the Office of Science and Technology Policy and Science and Technology Adviser to the President, made reference to a list of research questions submitted by the Cabinet at the request of the president. Selected examples of those questions were offered in the lecture, with a statement that the complete list would be made available.

Attached is the complete list.

Astronomy and Astrophysics

What is the nature of the universe? How did it originate? Is it expanding, contracting or in a steady state? How large and how old is it? [NASA]

Is there intelligent life elsewhere in the universe? [NASA]

What are the matter and energy mechanisms of stars—-quasars, pulsars, black holes? [NASA]

What is the nature of a solar flare? How is the energy stored and how is it released? [NASA]

How do planets evolve and what are the common processes that shape the environments of the Earth and the planets? [NASA]

How does the material pervading the universe collect to form complex organic molecules, stars, and galaxies? Research in this area can provide increased understanding of fundamental natural laws and the origins of the universe. [Press]

Biology and Microbiology

Can we discover anti-viral agents to combat viral diseases? The development of such drugs would have as large an effect on mankind as did the discovery of antibiotics. [DoD]

What are the mechanisms by which cells repair damage to their genetic material? This information will provide a better understanding of how the cells minimize mutations as a result of normal and imposed environmental stress. [DoE]

How do cells change during growth and development? Advances and understanding in this area should provide insights into the development of cell specialization and, perhaps, the aging process. [Press]

What are the molecular mechanisms by which genes are regulated to produce specialized products, and what new information is required to exploit the new DNA recombinant technology? This work may lead to improved knowledge of gene action. [Press]

Can microbiological research develop organisms which can convert crude organic materials, such as common cellulose, into livestock feed? The ability to convert common cellulose to feedstock would significantly increase the availability of high-grade animal protein for human consumption. [S]

What predisposing factors govern cellular differentiation and function in plants and animals? Successful research directed towards this question can provide an understanding in plants of factors responsible for drought tolerance and winter hardiness and in animals the mechanisms governing the development of fat and lean tissue. [DA]

What are the mechanisms by which hormonal substances regulate growth and reproduction in plants and animals? Answers to this vital question could help solve many perplexing problems, e.g., conception and embryonic mortality in animals and control of post-harvest ripening of fruits and vegetables. [DA]

In our eco-system affecting man and animals, how do microorganisms gain resistance to antimicrobial drugs and what mechanisms affect the maintenance and transfer of such resistance? Research to provide an understanding of bacterial resistance to drugs used in their control is essential for the protection of human and animal health. [DA]

What are mechanisms within body cells which provide immunity to disease? Research on how cell-mediated immunity strengthens and relates to other known mechanisms is needed to more adequately protect humans and animals from disease. [DA]

How can genetic improvement of crops for improved performance under stress conditions be accelerated? Research is needed to identify, more rapidly, useful gene sources for increasing photosynthetic efficiency and resistance to environmental stress. [DA]

What are the physical and biochemical factors associated with secondary cambial differentiation? The secondary cambium of a tree divides to form identical cells which are capable of becoming either phloem or xylem cells. Studies at the North Central Experiment Station are directed toward identifying the physical factors and biochemical signals which direct cambial development and differentiation. Such information will provide essential clues on the formation of wood. [DA]

How can utilization of the forest resource be enhanced through manipulations at the level of the plant cell, and through single-cell biodegradation? Tree cells can be stimulated to produce oleoresins, natural biocides, specific carbohydrates, and organic acids. Cell morphology such as fiber length can be altered to affect paper properties. Single-cell protein, hydrocarbons, acids, vitamins, steroids, and alcohols can be produced through biodegradation of tree components. [DA]

Can the microbiology of the gastrointestinal tract of man and animals be controlled? Research on this important question is needed to understand the contribution of microbial activity to general health and its effect upon nutrient utilization. [DA]

What are the quantitative differences between minimum human requirements for nutrients and those amounts needed for optimum physical, behavioral and mental functions? Research in this area will contribute to the attainment of maximum physical fitness and a longer, more vigorous, productive life. [DA]

Chemistry and Biochemistry

Combustion is older than recorded history, yet it is poorly understood in scientific terms. It is important that better understanding be achieved for all aspects of combustion, in order that our fossil fuels can be used with maximum efficiency and minimum adverse impact on the environment. [DoE]

To what extent can laser-induced chemistry be used as a practical, synthetic tool? Research in this area could lead to processes for preparing pure products with a low energy input and low environmental side effects. [DoE]

For many applications, solar energy is impractical because sunshine is intermittent, and energy storage is wasteful and expensive. Basic research is needed to develop ways in which sunlight can produce storable fuels. One possibility is to mimic but improve on photosynthetic processes, with emphasis on increased efficiency and products simpler than carbohydrates. Another approach is the use of sunlight to promote reactions which decompose water to hydrogen and oxygen. [DoE]

The liquefaction of coal is currently done by converting the complex coal structure to simple molecules, then recombining these into appropriate fuels. The process is capital intensive and energy wasteful. Research is needed on means to transform the coal into useful liquid fuels by a more direct route. This will involve much greater insight into the structure of coal and its reactions during the transformation process. [DoE]

How do catalysts work? Research on this question can lead to more economical ways to produce hydrogen and to convert coal to useful liquids and gases. [DoE]

What is the chemical basis of life? Where and how did it originate? Is a carbon-based chemistry a prerequisite for life? Does gravity play a significant role in the development and maintenance of life? [NASA]

Can simple chemical reactions be discovered that will generate visible radiation? The results of research on this question may lead to inexpensive lasers for communication and industrial uses. [Press]

Can new homogeneous catalysts be prepared that will catalyze chemical processes important to the chemical industry? Research in this area could make it possible to make specific molecules needed in industrial processing techniques with minimum energy expenditure and without the creation of unwanted molecules that may pollute the environment. [Press]

How do enzymes work? This research should help discover how enzymes selectively catalyze and control the chemical reactions carried out by living systems. The results of this research should extend knowledge on how to synthesize molecules in living cells. [Press]

What mechanisms of herbicidal action, at the cellular level, are responsible for weed-killing effectiveness? Understanding these mechanism is essential to improving technologies for reducing the $6 billion annual crop losses caused by weeds. [DA]

To what degree can conventional chemical pesticides be replaced by novel chemicals such as pheromones and insect growth regulators for forest insect pest suppression? Development of such chemicals would provide means of protecting the timber resource with minimal adverse environmental effects. [DA]

Earth, Oceans and Atmospheric Sciences

At what rate will atmospheric carbon dioxide concentrations increase as a result of increased use of fossil fuels? What effect will increasing carbon dioxide levels have on climate? How will this change the global social, economic and political structure? How might the impact be ameliorated? [DoE]

Can a predictive capability be developed regarding geochemical transport processes in the accessible regions of the earth's crust? Successful research directed toward this question would have major impact on expansion of the Nation's resource base, and would be of vital importance in resolving waste (nuclear and non-nuclear) problems. [DoE]

What is the nature of climate? What are the processes that control climate? How far into the future can you predict it? Is our climate warming or cooling? How far in advance can you predict weather, climate? Is there a relationship between climate and solar activity and, if so, what is the physical connection? [NASA]

What are the physical processes that govern climate? Greater understanding of climate could aid in the prediction of climate changes and allow time for measures to offset their impact. [Press]

To what extent is the stratospheric ozone affected by contamination of long-lived, man-made chemicals? The results of this research are important to man's survival and to the future of major industries. [Press]

What is the petroleum potential of the continental slopes and the adjacent ocean floors beneath deeper waters? This work is helping to identify the resource potential of the ocean's floor beyond the OCS. [Press]

How do organisms in the deep sea influence the productivity of the ocean? How will they react to sea floor dumping and mining activities? Answers to these questions will aid in assessing the future of the ocean as an important food source and should also provide baseline data on contamination of the sea. [Press]

Can research into the processes by which mineral deposits were formed in the earth's crust be sufficiently aided by deep ocean floor investigations so that mineral resources can be more efficiently located on land or sea-bed? Research which would improve the success-rate of exploratory efforts could be of considerable advantage. [S]

What improvement in understanding of oceanic and atmospheric effects on climate can be gained by increased use of sophisticated technology, such as satellites, in observing air/sea interactions? Air/sea interaction, is particularly Important in pursuing the promise of regional seasonal climate prediction and in determining the role of the ocean as the major absorber of atmospheric carbon dioxide (with implications for the fossil-fuel energy future). [S]

What physical processes govern the interaction between high energy plumes and the ambient atmosphere? Research in this area is needed to improve air pollution models and forest fire forecasts. [DA]

Economics

What is the economic and technical potential for saving energy in the processing and marketing of agricultural commodities? [DA]

What are the potentials for per capita energy saving and improved levels of living for alternative sizes and population densities of communities in the United States? [DA]

Despite continued long-term real economic growth in the United States, why are many rural areas chronically depressed? [DA]

What changes in policy at the Federal, state and local level can be designed to increase job opportunities in rural areas? A team research approach could provide a guide for changes in policy and more effective use of rural development funds. [DA]

What are the effects on farm income and consumer prices of environmental rules that pertain to farming? What environmental benefits result from such restrictions on farmers? [DA]

[47] What is the potential for microbial production of useful complex organic compounds including food products? Economic microbial processes for producing many complex organic chemicals from waste products appear feasible. [DA]

What are the individual and cumulative impacts of public domestic feeding programs on recipients and the Nation's economy? The annual level of current Federal programs is more than $7 billion. This research will facilitate analysis of alternative policy proposals. [DA]

How and how much is the instability of food and fiber product prices accelerating wage-price inflation and so handicapping real national economic growth? What gains in real economic growth would result from alternative price stabilizing mechanisms? What are the distributional effects of alternative economic gains and losses? [DA]

Since the production time frame for timber is much longer than for most agricultural crops, the economic consequences from trade policies in timber products may not be fully apparent for decades. Better economic methodologies are needed for assessing the gross national product, social welfare, and capital formation In developing countries. [DA]

Engineering, Computer and Material Sciences

Can man-machine interfaces be made so simple as to allow real time translation by untrained personnel. Such developments would not only provide improved communications between the nations, but also have a profound change in our daily life. [DoD]

How can productivity be enhanced by automation and artificial intelligence? With limited trained manpower supply in some areas and the saturation of productivity in others, it is extremely important for the nation to develop methods which will permit continual increases in productivity. [DoD]

Can new materials such as ceramics be developed to replace metals in high temperature situations. For example, use of ceramics in turbine blades will permit greatly increased operating temperatures, reduction of size, increase in efficiency, and reduction in fuel consumption. [DoD]

The economic and predictable fracture of rock is of critical importance to energy production. Obvious examples are drilling for new resources, mining of coal, oil shale and uranium, and releasing natural gas from low permeability formations. Research is needed on the mechanical behavior of rocks, in order to improve our understanding of them as engineering materials. [DoE]

Can new materials be developed which would be less dependent on critical or strategic elements? The obvious example of a benefit to be derived from research on this area is the possible substitution for Cr in steels. [DoE]

Computer models of physical and socio-economic processes are needed to guide, and often to replace, experimentation. Advances in analytical and numerical techniques and in computer hardware are required to simulate these processes more effectively. [DoE]

How do cracks initiate and propagate in materials? This research should provide information needed to develop structural materials that resist corrosion and failure under stress. [Press]

How can structures be designed and constructed to be both economical and earthquake resistant? In addition to reductions in life loss and personal injury, implementation of improved design procedures is expected to reduce losses to buildings alone by an average of $250 million per year. [Press]

How can the development of improved construction materials impact the cost of construction? Examples include soil stabilization and development of improved structural concretes. In FY 1976 and the Transition Quarter, DoT grants for construction totaled $7.86 billion. A 15% reduction in construction costs could have saved over $1 billion in this time period alone. [DoT]

What limitations, if any, exist on the development of full performance electric and hydrogen-powered automobiles, trucks and buses? areas of concern here include the development of improved batteries and lightweight, reliable and maintainable hydrogen storage tanks. The basic discipline area of interest is materials. The need for continued transportation energy research is obvious. [DoT]

What are the implications of the advent of inexpensive, large capacity microprocessors on the decentralized control of large scale systems in general and transportation networks in particular? I is necessary to expand our efforts in the area of modern control theory in order to understand the feasibility of and benefits from decentralized control. A situation now exists where the hardware state-of-the-art is advancing faster than our ability to employ it optimally. [DoT]

What are the long term impacts of major changes in the Nation's transportation system on the economic and social environment? How do improvements in transportation affect urban and regional form, the distribution of cities and the economic and production processes? Research on the development of formal methods and analytical tools such as large scale network analyses will be required to better understand the interaction of the major parameters and to more accurately predict the consequences of government policy decisions. [DoT]

Environmental and Ecological Sciences

Can specific bioprocessing methods be designed for removing and degrading toxic pollutants in industrial process and waste water? The benefits would be reduction of such agents to innocuous gases, production of chemical feed stock, and improvement of water quality. [DoE]

What are the ultimate carrying capacities of the terrestrial biosphere? [NASA]

What ecological factors and life-cycle phenomena govern insect dispersion and population explosions? Research on this question can lead to the development of innovative pest management technology to supplement current biological, cultural, and chemical control measures. [DA]

To what extent does nitrogen use in agriculture affect the ozone layer and what are the costs and environmental benefits of reduced nitrogen applications? [DA]

What is the chemical composition of precipitation and dry particulate matter and how does it vary with season and location? This information will provide baseline data for atmospheric input to nutrient cycling and can relate to both point and non-point sources of air pollution. [DA]

What factors influence susceptibility of harvested plant and animal products to post-harvest losses? In many parts of the world, such losses represent 30 to 50 percent of the food supply. Research can promote development of the technology required to preserve quality and protect against losses from rodents and insects. [DA]

How can the environmental stress tolerance of current crops and grasslands be improved? Utilization of as much as 40 percent of the world's noncultivated but potentially productive land is limited because of severe weather aberrations and other stress conditions. [DA]

General (Interdisciplinary)

To what extent can the occurrences of natural hazards such as fire, flood, earthquake, and pestilence be foreseen sufficiently in advance to permit mitigation of their effects? The problems of prediction and of mitigation are different for each hazard, but for each, research offers promise of reducing human and physical costs. [S]

What are the economic, technical, and public health impacts of restricting antibiotics and other additives in animal feeds? Adverse impacts may more than offset direct benefits of feed additive bans. [DA]

To what extent are agricultural chemicals transmitted to the Nation's waterways and what are the most cost-effective ways of reducing this pollution? [DA]

How will geoclimatic changes from increasing carbon dioxide levels and particulate loads in the atmosphere impact agricultural productivity? Research is needed to determine the influence of such changes on temperature, rainfall and other climatic variables that could significantly influence the agricultural potential of different regions. [DA]

What factors most influence the distribution of foods and so relate to human health? Research to help answer this question is needed prior to public nutrition programs. Such research could indirectly enable reduced health costs. [DA]

What are the economic potentials for expanding food production by (a) land and water development, and (b) application of new technology? This research would show the capability of the U.S. and other countries to meet the rising world demand for food. [DA]

Can productive self-sustaining systems be developed to utilize biological wastes? Research in this area could provide a means of improving comporting of organic wastes and preventing soil, water and air pollution with potential for yielding energy and other useful products. [DA]

Physics and Biophysics

Can materials be found that exhibit superconductivity at room temperature? Such a discovery would be extremely important to our energy needs as well as revolutionize all technology using electrical energy. [DoD and DoE]

Are there fundamental building blocks in nature? Some recent advances have been made which indicate that even the subnuclear "particles" are not fundamental and further research is necessary to uncover the secrets of the nucleus. [DoD]

How can considerations of second law efficiencies be incorporated into energy strategies? Energy should be valued not by its amount alone, but also by its thermodynamic quality. A significant reassessment of energy economics may be in order. [DoE]

How are the fundamental forces of nature related? Four types are currently known: nuclear (strong), electromagnetic, radioactive (weak) and gravitational. Only electromagnetism is well understood; the rest defy us to master them. [DoE]

Does an "island of stability" beyond the current periodic table or "abnormal" states of nuclear matter exist? These speculations can be tested and if found could have important consequences for nuclear energy production. [DoE]

What is the nature of gravity? Are there gravity waves, and if they exist, how do they propagate and at what velocity? [NASA]

What is the nature of matter? Why is matter and charge quantized? [NASA]

What are the limits for communications use of the channel capacity in the visible spectrum? Progress in this area could significantly expand the capacity of optical communication systems, and since these systems use glass fibers instead of copper, their use would result in tremendous monetary and resource savings. [Press]

Can microwave technology or other alternative sources of energy be safely and effectively used to process and preserve food? Food processing and preservation account for nearly 5% of the nation's consumption of fossil energy. Research could provide alternative less costly energy sources and methodology. [DA]

Social, Psychological and Behavioral Sciences

What is the nature of intelligence [NASA]

How do we think? [NASA]

What are the individual and cumulative effects of government regulation on productivity? This research will provide a sound technical basis for assessing the benefits and cost of proposed, as well as existing, government regulations. [Press]

What are the factors controlling cognitive development? For example, how can the large number of component processes involved in reading and understanding a paragraph be characterized? Research on this question should provide new knowledge on the processes involved in reading and comprehending text. Such work is important on providing a basis for improving the techniques for teaching people to read and comprehend. [Press]

What are the mechanisms responsible for sensory signal processing, neural membrane phenomena, and distinct chemical operations of nerve junctions? Research in these areas will extend knowledge of perception, behavior, and the chemical functioning of the nervous system. [Press]

What are the factors—social, economic, political, and cultural—which govern population growth? High population growth rates in the developing countries impose an economic burden which too often exceeds the gains made by development. Social and biomedical research on safe, efficacious, and culturally acceptable contraceptives would therefore be of great benefit. [S]

The following item was in Press's original list but not in the master list.

To what degree can biological nitrogen fixation be enhanced? Successful research directed toward this question may provide more information on joint plant-bacteria relationships and an environmentally sound method of increasing crop productivity while minimizing energy costs. [Press]

Appendix D
Profiles of Scientists' Voluntary Public-Interest Associations

This compilation of profiles of scientists' public-interest associations and related organizations is intended to increase awareness of an important yet little-known and underappreciated interface between science and society.

The profiles are based on the sources referenced and on information found in the associations' own Web pages. (Quotations not attributed to other sources are from Web pages.)

It was our standard procedure to send drafts to organizations for comments and corrections. We prepared the profiles carefully, but we cannot guarantee their correctness.

The organizations listed here in alphabetical order should be considered exemplars. Especially in peripheral categories, the compilation is far from exhaustive.

Some of the information (most of which was gathered in 1999 and 2000) is going to be obsolete quickly. It appears to be technically feasible to maintain an updated electronic database of these associations on the Internet, and we would be happy to provide our data files for such a venture.

We hope to stimulate further research by others, on their own or in collaboration with us. Additional material, updates, revisions, etc., will be gladly received, to be considered for use in the database and in future editions of this volume.

The following organizations are profiled here:

American Association of Scientific Workers
American Indian Science and Engineering Society
Asilomar Conferences
Association for Women in Science
Association of Scientific Workers
Atomic Scientists' Association
Bulletin of the Atomic Scientists
Center for Democracy and Technology
Center for Science in the Public Interest
Citizens' League Against the Sonic Boom
Civilian R&D Foundation for the Independent States of the FSU
Coalition for Responsible Genetic Research
Committee for Nuclear Responsibility

Committee of Concerned Scientists
Committee on Human Rights of Scientists (New York Academy of Sciences)
Committee on International Security Studies (American Academy of Arts and Sciences)
Committee on Science, Engineering, and Public Policy (National Academy of Sciences)
Committee on the International Freedom of Scientists (American Physical Society)
Computer People for Peace
Computer Professionals for Social Responsibility
Conservation Law Foundation
Council for a Livable World
Council for Responsible Genetics
Creation Research Society
Education Development Center
Emergency Committee of Atomic Scientists
Environmental Defense Fund (later Environmental Defense)
Environmental Law Institute
Environmental Literacy Council
Ethical, Social and Legal Implications (Human Genome Project)
Ethics and Values in Science and Technology (now Societal Dimensions of Engineering, Science, and Technology) program (National Science Foundation)
Federation of American Scientists
Forum on Physics and Society (American Physical Society)
General Education in Engineering
George C. Marshall Institute
Hastings Center
Hudson Institute
Institute for Science and International Security
Institute on Religion in an Age of Science
Intergovernmental Panel on Climate Change
International Council for Science
International Dark-Sky Association
International Institute for Applied Systems Analysis
International Network of Engineers and Scientists for Global Responsibility
International Peace Research Institute, Oslo
International Physicians for the Prevention of Nuclear War
International Science and Technology Center
Loka Institute
Médecins Sans Frontières
Medical Committee for Human Rights
National Action Council for Minorities in Engineering
National Center for Science Education
National Conference of Lawyers and Scientists

National Society of Black Engineers
National Women's Health Resource Center
Natural Resources Defense Council
New University Conference
NIH Black Scientists Association
Office on Public Understanding of Science (National Academy of Sciences)
Peace Action (formerly SANE/Freeze)
Physicians for a National Health Program
Physicians for Human Rights
Physicians for Social Responsibility
Pugwash Conferences
Red Crate Collective
Science and Environmental Policy Project
Science and Human Rights Program (American Association for the Advancement of Science)
Science in a Social Context
Scientific Freedom, Responsibility and Law Program (American Association for the Advancement of Science)
Scientists and Engineers for Social and Political Action / Science for the People
Scientists for Global Responsibility
Scientists for Orlov and Shcharansky (later Scientists for Sakharov, Orlov, and Shcharansky)
Scientists' Institute for Public Information
Society for Advancement of Chicanos and Native Americans in Science
Society for Community Research and Action
Society for Freedom in Science
Society for Social Responsibility in Science
Society for the Psychological Study of Social Issues
Society for the Study of Peace, Conflict and Violence
Society of Hispanic Professional Engineers
Standing Committee on Scientific Freedom and Responsibility (American Association for the Advancement of Science)
Stockholm International Peace Research Institute
Union of Concerned Scientists
Women in Engineering Programs and Advocates Network
Working Group on Ethical, Social and Legal Implications (Human Genome Project)
World Federation of Scientific Workers

American Association of Scientific Workers

With the left-leaning British Association of Scientific Workers—in existence since 1918—as a model, an American counterpart was organized at a meeting of the American Association for the Advancement of Science in Richmond, Virginia on

December 30, 1938. The roots of AAScW were a Philadelphia group led by K. A. C. Elliott and a Boston-Cambridge group initiated by Kenneth V. Thimann. Among the original sponsors of AAScW were Harold Urey, A. J. Carlson, Robert Chambers, A. C. Ivy, and Henry Sigerist.

Purpose AAScW's goals were to "orient scientific developments to their social implications, and give scientists a measure of control over the applications of science" (Strickland 1968: 13).

History AAScW attracted prominent sponsors, such as Arthur Compton, Karl Compton, and J. Robert Oppenheimer. In the spring of 1939, the Boston-Cambridge chapter organized a boycott of products from Hitler's Germany. Yet AAScW soon changed course and, in the spring of 1940, submitted a neutralist petition, signed by 500 scientists, to President Roosevelt. This caused a group of eminent Princeton scientists, among them Albert Einstein and Eugene Wigner, to communicate their disapproval of the petition to the president, and it also prompted a counter-petition, with Linus Pauling among its signatories. There were rumors and accusations that AAScW was a kind of communist front organization. In protest against the alleged communist control over the Boston-Cambridge branch, several members openly resigned, among them G. B. Kistiakowsky and George Wald. Immediately at the end of World War II, AAScW, under the leadership of the geologist Kirtley Mather, spoke out on the issue of nuclear weapons before the Manhatten Project scientists themselves had formed public-interest associations. AAScW virtually disappeared soon thereafter.

References

Nichols, D. 1974. The associational interest groups of American science. In *Scientists and Public Affairs*, ed. A. Teich. MIT Press.

Kuznick, P. 1987. *Beyond the Laboratory*. University of Chicago Press.

Smith, A. 1965. *A Peril and a Hope*. University of Chicago Press.

Strickland, D. 1968. Scientists in Politics. Purdue University Studies.

Wang, J. 1999. *American Science in an Age of Anxiety* University of North Carolina Press.

American Indian Science and Engineering Society

Concerned about the extremely high high-school dropout rates of American Indians, about their low college enrollment and graduation rates, and about the severe underrepresentation of American Indians in science and engineering, American Indian scientists, engineers, and educators founded AISES in 1977.

Purpose "Through its educational programs, AISES provides opportunities for American Indians and Native Alaskans to pursue studies in science, engineering, business and other academic arenas. The trained professionals then become technologically informed leaders within the Indian community. AISES' ultimate goal is to be a catalyst for the advancement of American Indians and Native Alaskans as they seek to become self-reliant and self-determined members of society."

Current status and activities AISES is a charitable nonprofit organization, with its national headquarters located in Albuquerque, NM. Currently there are about

3,500 members and 63 affiliated high schools. AISES has professional chapters and college chapters. The student members elect representatives to the Society's governing boards as well as their own national, regional and campus officers. "Through a variety of educational programs, AISES offers financial, academic and cultural support to American Indians and Alaska Natives from middle school through graduate school." In the 1998/99 academic year, it awarded more than $600,000 in scholarships at the undergraduate and graduate levels. AISES organizes a National Science Fair at which American Indian students exhibit original scientific research projects, and an annual conference. The Society also provides professional development activities to enable teachers to work effectively with Native American students. It issues the quarterly *Winds of Change* and other publications, among them culturally appropriate curricula.

Web address http://www.aises.org

Mailing address P.O. Box 9828, Albuquerque, NM 87119-9828

Phone number 505 765 1052

Fax number 505 765 5608

E-mail address info@aises.org

Asilomar Conferences

Important conferences were held in 1973 and 1975 at the Asilomar Conference Center in Pacific Grove, California. The 1973 conference was organized by the biochemist Paul Berg (Stanford) with Robert Pollack (Cold Spring Harbor Laboratory), Michael Oxman (Harvard Medical School), and Al Hellman (NCI). It was supported by NSF and NCI and had about 100 participants (all from the US). The purpose was to discuss the problem of biohazards of research involving tumor viruses. A major issue was that of informed consent of laboratory workers and surrounding communities. Among the findings and recommendations of the conference were that neither safety nor danger of the materials in question had been proved; that the laboratory community should be the subject of epidemiological studies; that a major commitment for safety equipment must be made, as well as a commitment to the safety education of laboratory personnel; that a registry of health indices and laboratory experiences of laboratory personnel should be established; and that informed-consent statements should be required from grantees and employees (Krimsky 1982: 63). At the Gordon Conference on Nucleic Acids in 1973, a majority of participants supported a letter that was sent to the president of the National Academy of Sciences and published in *Science*. The letter described, among other things, some potential risks of recombinant DNA techniques, and recommended establishing a study committee (Krimsky 1982: 78). In July 1974, the resulting panel of leading researchers in the field distinguished three classes of experiments in recombinant DNA research and published a call for a voluntary moratorium of the first two classes and a careful approach to the third. They also called for an NIH committee to assess the risks involved and to establish guidelines. The 1975 conference included foreign participants. Its purpose was to assess the risks of the new technology and to issue recommendations that would establish the basis for the NIH guidelines. These guidelines, issued 16 months later, were indeed largely patterned after the conference recommendations. The second Asilomar conference thus played a crucial role in shaping rDNA policy for decades. The whole process serves as a major example of self-regulation within the scientific community.

References

Krimsky, S. 1982. *Genetic Alchemy*. MIT Press.

Weiner, C. 1979. The recombinant DNA controversy: Archival and oral history resources. *Science, Technology, and Human Values* 26: 17–19.

Association for Women in Science

Founded in 1971, the AWIS has as its goals equity and full participation for women in science, mathematics, and technology. A major focus has been mentoring. In a multi-year effort, AWIS established and improved community mentoring programs at the undergraduate and graduate student levels. Another major project studied the academic climate for women science faculty through site visits and evaluations, and issued recommendations.

Current status and activities AWIS's headquarters are located in Washington, with a small professional staff (and interns); at the local level, there are 76 chapters around the US. The association has more than 5,000 members. AWIS continues to conduct projects on issues relating to women in science and engineering, and is currently creating a comprehensive database of women scientists and engineers. AWIS also organizes conferences and publishes a variety of materials, among them the bimonthly *AWIS Magazine*. It tries to shape national policy through Congressional testimony and participation in various national coalitions. The AWIS Educational Foundation offers several awards and scholarships.

Web address http://www.awis.org

Mailing address 1200 New York Avenue NW, Suite 650, Washington, DC 20005

Phone number 202 326 8960

Fax number 202 326 8960

E-mail address awis@awis.org

Association of Scientific Workers

This association was founded in London in 1918 to satisfy the need of scientists to express their views independently, apart from advising the government.

Purpose To foster a new and more vivid awareness of the need for the social control and social use of science and of the dangers which threaten if it is not so controlled and used.

History At the end of World War II, AscW had 16,000 members, among them many prominent scientists. AscW saw itself as a scientists' and scientific workers' trade union and was affiliated with the Trades Union Congress. In that role, it sought to protect and advance the interests of its members, as well as to promote its more general aim. AscW was absorbed into the Association of Scientific,

Technical and Managerial Staff (ASTMS) trade union, which later merged into the Manufacturing, Science and Finance union.

Reference Smith, A. 1965. *A Peril and a Hope*. University of Chicago Press.

Atomic Scientists' Association

A group of physicists, chemists, and engineers, many of whom had participated in the Atomic Bomb project, formed ASA to influence Britain's policies on nuclear weapons and atomic power. The origins of this group lay in a committee on atomic energy of the British Association of Scientific Workers, which separated from that organization and founded ASA as an independent association on March 8, 1946 in London. One of the co-founders was Joseph Rotblat.

Purpose To bring before the public of Britain the facts about atomic energy and its implications, to investigate and make proposals for the international control of atomic energy in order to help in the solution of the most pressing problem, and to help shape the policy of Britain in all matters relating to atomic energy.

History The ASA consisted of a National Council and Regional Branches, each based on a university. The branches arranged public meetings and provided speakers to schools, trade unions, workers' educational associations, co-operative societies, etc. The national organization built the "Atom Train," an exhibition that toured Britain in two railway coaches. A guide to the exhibition sold more than 100,000 copies. ASA issued a monthly bulletin, *Atomic Scientists' News*, and produced a leaflet, *Atomic Survey*, which provided a short explanation of atomic and nuclear physics. Full membership was "limited to graduate scientists with specialized knowledge of Atomic Energy; the Association can thus speak as a body of experts." Associate membership was "open to all interested members of the public."

Reference Smith, A. 1965. *A Peril and a Hope*. University of Chicago Press.

Bulletin of the Atomic Scientists

This publication was conceived by Hyman Goldsmith, Eugene Rabinowitch, and Edward Shils (Smith 1965: 294). The first issue appeared on December 10, 1945. The *Bulletin* was originally intended to provide a forum for the Chicago group of nuclear scientists (Atomic Scientists of Chicago) and for FAS issues. Initially titled *Bulletin of the Atomic Scientists of Chicago*, it soon acquired a national outlook and dropped "of Chicago."

Purpose To warn of the dangers of nuclear war, and to work toward some form of international control of atomic weapons.

History By 1947, the *Bulletin* had a circulation of about 20,000 in 17 countries (Gottfried 1999: 43). Long-time editor and driving force of this publication has been Eugene Rabinowitch (Moot 1986). The *Bulletin* has been the major voice of the scientists' movement against nuclear arms.

Current status and activities The bi-monthly *Bulletin* is published by the not-for-profit Educational Foundation for Nuclear Science. The content is described as "nuclear plus"—maintaining the special focus on nuclear issues, but also covering a wider range of national and international security issues. A well-known feature of the *Bulletin* is the Doomsday Clock to dramatize the nuclear perils.

Web address http://www.bullatomsci.org

References

Gottfried, K. Physicists in politics. *Physics Today*, March 1999: 42–48.

Moot, A. 1986. Eugene Rabinowitch and the *Bulletin of the Atomic Scientists*: Responding to Responsibility. Senior thesis, Harvard University.

Smith, A. 1965. *A Peril and a Hope*. University of Chicago Press.

Center for Democracy and Technology

Purpose "To promote democratic values and constitutional liberties in the digital age. With expertise in law, technology, and policy, CDT seeks practical solutions to enhance free expression and privacy in global communications technologies. CDT is dedicated to building consensus among all parties interested in the future of the Internet and other new communications media." CDT believes that "the open, decentralized, user-controlled, and shared resource nature of the Internet creates unprecedented opportunities for enhancing democracy and civil liberties," and wishes to seek "public policy solutions that preserve these unique qualities and thereby maximize the democratizing potential of the Internet." Guiding principles include the belief in the freedom of expression on the Internet, the recognition of the importance of Internet privacy, and the goal of fostering widely available, affordable Internet access.

Current status and activities CDT is a 501(c)(3) non-profit public policy organization, composed of diverse working groups of public-interest and commercial representatives who conduct research, are involved in public education campaigns, litigation, public policy advocacy, and the development of technology standards and on-line information resources. In 1998, 33 firms, associations, and foundations from all areas of communications and computer industries (including Microsoft, IBM, Time Warner, and the Open Society) funded CDT's working groups and special project activities. CDT provides web resources to educate the public about current policy issues concerning the Internet, such as the "CDT Guide to On-line Privacy," the "CDT Privacy Quiz," Policy Posts, and an updated guide to "Legislation Affecting the Internet." CDT also seeks to share its knowledge with members of Congress through the Congressional Internet Caucus and the Internet Education Foundation. Other efforts include mobilizing grassroots participation (such as organizing the Citizens Internet Empowerment Coalition [CIEC], a diverse group of Internet users, librarians, publishers, on-line service providers, and civil liberties groups), and coordinating a series of working groups. These groups serve as forums for communications service providers, computer hardware and software producers, content providers, consumer and privacy advocates, and other non-profits to exchange knowledge and find solutions to policy problems in areas of on-line privacy, digital security, and free expression. CDT participates in international activism; for example, through the Global Internet Liberty Campaign, it co-sponsors a conference of human rights organizations in Budapest. It also produced the report "Regardless of Frontiers: Protecting the Human Right to Freedom of Expression on the Global Internet."

Web address http://www.cdt.org

Mailing address 1634 Eye Street, NW, Suite 1100, Washington, DC 20006

Phone number 202 637 9800

Fax number 202 637 0968

E-mail address webmaster@cdt.org

Center for Science in the Public Interest

Founded in 1971, in Washington, by Albert J. Fritsch (a chemist), Michael F. Jacobson (a microbiologist), James B. Sullivan (a meteorologist), and Kenneth Lasson (an attorney). The scientists had met a year earlier at Ralph Nader's Center for the Study of Responsive Law. Part of the founders' motivation was to serve as role models for other scientists to become more active in public-interest endeavors.

Purpose "Focuses on improving the safety and nutritional quality of our food supply and on reducing the carnage caused by alcoholic beverages. CSPI seeks to promote health through educating the public about nutrition and alcohol; it represents citizens' interests before legislative, regulatory, and judicial bodies; and it works to ensure that advances in science are used for the public's good."

History During the organization's low-budget early years, the three scientist-founders and co-directors each followed their own interests that were grounded in their respective scientific backgrounds. Fritsch did projects on toxic chemicals, strip mining, nuclear power, and oil, Jacobson worked on food additives and nutrition, and Sullivan focused on highways and air pollution. When, in 1977, two of the original three scientists left, this had two important consequences. CSPI now focused almost exclusively on food issues, the specialty of Michael Jacobson, the remaining scientist. Secondly, at the organizational level, Jacobson became executive director, and CSPI was set on the route toward professionalization. From then on, the staff has grown from 15 to 55, and the annual budget from less than $1 million to more than $14 million. Starting with the first Food Day in 1975, a long campaign to combat the disease-promoting effects of modern-day diet and to curb deceptive labeling by food manufacturers, successfully culminated in the passing of the Nutrition Labeling and Education Act of 1990. CSPI has also been concerned with restaurant food. In 1986, the first edition of the *Fast Food Guide* provided detailed nutritional information on the entire menu lines of popular fast food chains. Another focus of CSPI has been alcohol. In 1982, CSPI started its program on alcoholic beverages, advocating higher taxes and less advertising.

Current status and activities CSPI is a nonprofit education and advocacy organization, headquartered in Washington, with a staff of 50. CSPI has nearly a million subscribers to its *Nutrition Action Healthletter*, the largest-circulation health newsletter in the US. In 2000, subscription revenue provided about 75 percent of CSPI's annual budget of $14 million. Apart from the *Healthletter*, CSPI produces various publications and disseminates a great deal of health and nutrition-related information on its Web site.

Web address http://www.cspinet.org

Mailing address 1875 Connecticut Avenue, NW, Suite 300, Washington, DC 20009

Phone number 202 332 9110

Fax number 202 265 4954

E-mail address cspi@cspinet.org

Citizens' League Against the Sonic Boom

Founded in 1970, this group campaigned and lobbied against plans to build a supersonic transport plane in 1970–71.

History CLASB had 5,000 members, half of whom were high school and college students. It joined with other public-interest associations to form the *Coalition Against the SST*. CLASB's director was Harvard physicist W. A. Shurcliff, the associate director Harvard biologist J. T. Edsall. The headquarters were located in Cambridge, Massachusetts. This single-issue group dissolved after the primary purpose was accomplished.

References

Hedal, L. 1980. Citizen's League Against the Sonic Boom vs. the National Academy of Sciences: Scientific Activists versus Scientific Advisors. A.B. thesis, Harvard University.

Krimsky, S. 1982. *Genetic Alchemy*. MIT Press.

Nichols, D. 1974. The associational interest groups of American science. In *Scientists and Public Affairs*, ed. A. Teich. MIT Press.

Shurcliff, W. 1970. *S/S/T and Sonic Boom Handbook*. Ballantine.

Civilian R&D Foundation for the Independent States of the FSU

In 1995, in response to the decline of science and technology in the former Soviet Union (FSU), the US government set up CRDF as a private, non-profit charitable organization. The 1992 legislation that authorized the creation of this foundation was sponsored by Representative George Brown and Senators Albert Gore and Joseph Lieberman.

Purpose To foster scientific and technological collaboration between US and FSU researchers, and to encourage the growth of civilian employment opportunities for former FSU defense scientists; to help prevent the dissolution of the scientific infrastructure in the FSU; to advance defense conversion; to assist in the establishment of a market economy in the FSU.

History NSF, which the legislation directed to establish CRDF, transmitted initial funds of about $10 million to CRDF in an Endowment Agreement and appointed its directors.

Current status and activities Each participating FSU country must provide a minimum amount of financial support for CRDF. The Foundation conducts a great variety of programs, under the major headings of Research Grant Programs, Industry-Oriented Programs, and Institution-Building Initiatives.

Web address http://www.crdf.org

Mailing address 1800 N. Kent Street, Suite 1106, Arlington, VA 22209

Phone number 703 526 9720

Fax number 703 526 9721

E-mail address information@crdf.org

Coalition for Responsible Genetic Research

Debates about the safety of genetic experiments led to hearings before the City Council of Cambridge, Massachusetts. In 1976, in the aftermath of these hearings, CRGR was formed as a coalition of scientists and environmentalists. George Wald and Lewis Mumford were among the initial members.

History Together with like-minded public-interest groups, CRGR lobbied for strict regulations of genetic research. Among its activities was a mass letter-writing campaign.

Reference Krimsky, S. 1982. *Genetic Alchemy*. MIT Press.

Committee for Nuclear Responsibility

The driving force behind CNR has been John William Gofman, now Professor Emeritus of Molecular and Cell Biology at the University of California at Berkeley. As a graduate student and postdoctoral fellow, Gofman had been involved in research on plutonium and in the Manhattan Project. In 1963, he established the Biomedical Research Division at the Lawrence Livermore National Laboratory to evaluate the health effects of all kinds of nuclear activities. Research in this area led Gofman and Arthur Tamplin to conclude, in 1970, that radiation was far more carcinogenic than admitted by the Atomic Energy Commission. They publicly called for a five-year moratorium in licensing additional nuclear power plants (and lost their research grants as a consequence). Their concern with the health risks of radiation led to the formation of CNR in 1971.

Purpose To "provide independent analyses of sources and health effects of x-rays and other ionizing radiations." CNR works toward preventing careless x-ray overdosing as well as additional nuclear pollution, with the goal of reducing human cancer and other adverse health effects.

History In 1992, Gofman shared the Right Livelihood Award (which is known as "the Alternative Nobel Prize") for "his pioneering work in exposing the health effects of low-level radiation."

Current status and activities A nonprofit educational group, CNR publishes short non-technical and technical papers and longer books, such as *Radiation-Induced Cancer from Low-Dose Exposure* (1990) and *Preventing Breast Cancer* (1996), both by John W. Gofman.

Web address http://www.ratical.org/radiation/CNR/

Mailing address P.O. Box 421993, San Francisco, CA 94142

Phone number 415 776 8299

Fax number 415 776 8299

E-mail address cnr123@web.tv

Committee of Concerned Scientists

Founded in 1972 in Washington and New York, this organization began as an ad hoc group of scientists, engineers, and physicians. They were motivated by a concern for colleagues whose fundamental personal and scientific rights were denied, and they felt that constructive nationwide action should be taken on the oppressed scholars' behalf.

Purpose "The protection and advancement of the human rights and scientific freedom of colleagues around the world."

History Originally CCS focused on helping oppressed Soviet scholars, such as Andrei Sakharov, Yuri Orlov, Benjamin Levich, Anatoly Shcharansky, and other scientific dissidents. Before long, this focus on the former Soviet Union was expanded to include violations of the human rights and scientific freedom of colleagues around the world. In 1999, CCS supported scientists in 21 countries.

Current status and activities CCS monitors and documents violations of the human rights and scientific freedom of scientists, engineers, and physicians worldwide. It publicizes the plight of persecuted colleagues, protests directly to repressive governments and/or offending employers, and works to secure for these colleagues the support of learned societies and the US government. It also organizes visits and scientific exchanges involving endangered colleagues, and provides material assistance, as needed, to them and their families in cases when they have been dismissed from their jobs or jailed.

Web address http://www.libertynet.org/ccs

Mailing address 53-34 208th Street, Bayside, NY 11364

Phone number 718 229 2813

Fax number 718 229 7540

E-mail address ccs@cims.nyu.edu

Committee on Human Rights of Scientists (New York Academy of Sciences)

In 1978 a standing committee of the New York Academy of Sciences was formed to support the basic human rights of scientists throughout the world.

Purpose The Committee intervenes in cases where scientists, engineers, health professionals, and educators are detained, imprisoned, exiled, murdered, "disappeared," or deprived of the rights to pursue science, to communicate their findings with their peers and the general public, and to travel freely in accordance with established rules of the International Council of Scientific Unions. The Committee's means of intervention include letters to those in authority, personal meetings and appeals, on-the-scene and behind-the-scene pressure on American and foreign authorities, public statements, and petitions.

Current status and activities The Committee has closely coordinated its activities with other major human rights groups. It participates jointly in campaigns with these groups, while maintaining its independence and identity as a standing committee of the New York Academy of Sciences.

Reference Committee on Human Rights of Scientists, Mission and Mandate Statement.

Web address http://www.nyas.org

Mailing address 2 East 63rd Street, New York, NY 10021

Phone number 212 838 0230

Fax number 212 838 5226

E-mail address nyas@nyas.org

Committee on International Security Studies (American Academy of Arts and Sciences)

The Academy formed CISS in 1982 to formalize and expand its work on international security affairs. The Academy's interest in this area had begun much earlier, with a 1960 special issue of *Daedalus* on Arms Control and with its sponsorship of the US Pugwash Committee since the early 1960s.

Purpose To study problems in the broad area of international security.

Current status and activities With a membership of 24 persons, CISS oversees the US Pugwash Committee, sponsors public events, and publishes project reports, occasional papers, and books. Current and recent CISS projects have studied topics such as the International Criminal Court and US national security; small arms and light weapons, environmental scarcities, state capacity and civil violence; security in the post-Soviet space; emerging norms of justified intervention; and the Israeli-Palestinian peace process.

Web address http://www.amacad.org/projects/ciss.htm

Mailing address 136 Irving Street, Cambridge, MA 02138

Phone number 617 576 5024

Fax number 617 576 5050

E-mail address tsanderson@amacad.org

Committee on Science, Engineering, and Public Policy (National Academy of Sciences)

In the aftermath of the "Sputnik shock," Presidential Science Advisers George Kistiakowsky and Jerome Wiesner felt that the National Academy of Sciences could provide a vehicle for tapping the expertise of the US research community on broad policy matters. In 1961, the NAS Committee on Government Relations was founded; in 1962, the committee received its first grant from the National Science Foundation.

Purpose COSEPUP's formal responsibility is "to deliberate on initiatives for new studies in the area of science and technology policy, taking especially into account the concerns and requests of the President's Science Advisor, the Director of the NSF, the Chairman of the National Science Board, and the chairmen of key science and technology-related committees of Congress." In conducts "studies on cross-cutting issues in science and technology policy." Studies are carried out by special interdisciplinary panels.

History In 1963, the name changed from Committee on Government Relations to Committee on Science and Public Policy (COSPUP). In 1982, the policy-related activities of the National Academy of Engineering were merged with COSPUP, thus creating COSEPUP. The currently existing COSEPUP has been jointly chartered by the National Academy of Sciences, the National Academy of Engineering, and the Institute of Medicine. At this writing, COSEPUP has issued 49 documents on issues such as the federal support of basic research, technology assessment and choice, global warming, scientific conduct, and information technologies. It should be noted that other committees and units of the National Academy of Sciences are also involved with issues at the interface of science, society, and politics (Halpern 1997).

Current status and activities Recently completed COSEPUP reports include "An Assessment of the National Science Foundation's Science and Technology Centers Program," "Careers in Science and Engineering: A Student Planning Guide to Grad School and Beyond," "Reshaping the Graduate Education of Scientists and Engineers," and "Advisor, Teacher, Role Model, Friend" (all available on line).

Reference Halpern, J. The US National Academy of Sciences—In service to science and society. *Proceedings of the National Academy of* Sciences 94 (1997): 1606–1608.

Web address http://www2.nas.edu/cosepup/

Mailing address COSEPUP, National Academy of Sciences, 2101 Constitution Avenue, NW, Washington, DC 20418

Phone number 202 334 2424

Fax number 202 334 1667

E-mail address cosepup@nas.edu

Committee on the International Freedom of Scientists (American Physical Society)

Purpose According to the APS bylaws, CIFS "shall be responsible for monitoring concerns regarding human rights for scientists throughout the world. It shall apprise the President [of APS], the Executive Board and Council of problems encountered by scientists in the pursuit of their scientific interests or in effecting satisfactory communication with other scientists."

History CIFS has helped persecuted scientists, and especially physicists, throughout the world. Since the end of the Cold War, the major focus for CIFS's work has been the People's Republic of China.

Current Status and activities CIFS is a standing committee of APS. Usually in collaboration with the Committee of Concerned Scientists and the Committee on the Human Rights of Scientists of the New York Academy of Sciences, CIFS sponsors petitions on behalf of scientists whose rights have been violated. It also organizes sessions at conferences.

Web address http://www.aps.org/intaff/cifs1.html

Mailing address One Physics Ellipse, College Park, MD 20740

Phone number 301 209 3200

Fax number 301 209 0865

E-mail address irwin@aps.org

Computer People for Peace

Founded in 1968 as Computer Professionals for Peace, this group had about 200 members and an informal steering committee located in New York. CPP's outlook was radical. One of its activities in 1969 was to collect bail money for the defendants of the Black Panther 21 trial in New York. CPP also issued a newsletter titled *Interrupt*. The group is now defunct.

Reference Nichols, D. 1974. The associational interest groups of American science. In *Scientists and Public Affairs*, ed. A. Teich. MIT Press.

Computer Professionals for Social Responsibility

In October 1981 a discussion group formed among computer researchers at the Xerox Palo Alto Research Center who were worried about the threat of nuclear war and wanted to integrate their work life with their social concerns. CPSR grew out of this informal discussion group. It was incorporated in 1983. The first national chairperson was Severo Ornstein and the first national secretary was Laura Gould. Gary Chapman was hired as executive director in 1985.

Purpose "To provide the computer science and the general public with scientific information and expert judgment on which social and political decisions must in part be based." Among the original foci of CPRS's attention were the threats and dangers posed by the nuclear arms race and the computer technology associated with it.

History President Reagan's Strategic Defense Initiative (SDI), which was announced two weeks after CPRS's incorporation, immediately became the principal target of CPRS's activity. CPRS's campaign against SDI led to the opening of chapters in various parts of the US and boosted membership numbers. CPSR spread its message through publications, conferences, and special events. Since the mid 1980s, the scope of CPSR's program has widened considerably to include a variety of social issues related to computers. In 1986, for instance, a privacy and civil liberties project was launched, as well as a computers in the workplace project. A Washington office opened in 1988 to house the privacy initiative. This office became independent as the Electronic Privacy Information Center (EPIC) in 1994. Between 1991 and 1993, an office in Cambridge, Massachusetts, focused on reorienting science and technology toward non-military solutions of national and international problems. In March 1993, CPSR started a project on the National Information Infrastructure (NII) proposed by the Clinton administration.

Current status and activities CPSR, which is tax-exempt, has 24 chapters across the US and a national structure headed by an executive director, a board of directors, and an advisory board. To focus on specific issues, members' working groups are instituted that pursue their activities with a high degree of autonomy. Currently five working groups exist (Cyber-Rights, Civil Liberties, Workplace, Y2K [the year 2000 computer problem], Education). Two additional working groups (Ethics, Law) are in the process of formation. CPRS gives out the Norbert Wiener Award for excellence in promoting the responsible use of technology.

Reference Johnson, J. Conflicts and Arguments in CPSR [on Web site]

Web address http://www.cpsr.org/

Mailing address P. O. Box 717, Palo Alto, CA 94302

E-mail address cpsr@cpsr.org

Conservation Law Foundation

Partly in response to attempts to develop slopes of Mount Greylock, the highest peak in Massachusetts, seven prominent New England environmentalists founded CLF on October 6, 1966.

Purpose The CLF works "to solve the environmental problems that threaten the people, natural resources and communities of New England. CLF's advocates use law, economics and science to design and implement strategies that conserve natural resources, protect public health, and promote vital communities in our region."

History A first legal success came in 1967, when the Massachusetts Supreme Judicial Court agreed with a friend-of-the-court brief filed by CLF and blocked plans to widen Route 2 by filling part of Spy Pond in Arlington. In 1972 the first paid staffer was hired. In the 1970s, CLF's main focus was on crafting a host of environmental regulations, especially on land use. In 1977, CLF helped avert the building of a four-lane divided highway through the scenic landscape at Franconia Notch, New Hampshire. In the following years, CLF sued to prevent off-shore oil drilling at Georges Bank and to stop construction of an oil refinery in Eastport, Maine. Both actions ended in success. In 1979, the first staff scientist was hired, Linzee Weld. In 1983, CLF initiated a court case that resulted in the huge Boston harbor clean-up project. Starting in the mid 1980s, a CLF campaign highlighted the dangers of lead pollutants to the health of citizens, especially of young children. In 1988, Massachusetts passed the toughest law in the nation to prevent lead poisoning.

Current status and activities CFL is a nonprofit, member-supported organization that is active in the New England area. CLF has 45 staff members, among them about 20 attorneys, in four regional advocacy centers (Boston, Massachusetts; Montpelier, Vermont; Rockland, Maine; Concord, New Hampshire). The Boston office is also in charge of membership, development, and communications. CLF offers summer jobs, internships, and volunteer opportunities. In 1998, it had an annual organizational budget of more than $3.5 million. CLF issues a quarterly magazine titled *Conservation Matters* and a variety of other publications. In addition to its own legal actions, CLF also provides legal representation to other environmental groups. In its Community Advocacy Initiative, CLF joins forces with local citizens' groups. A recent addition to the spectrum of CLF activities is entrepreneurial environmentalism, or environmental business transactions called "Ventures."

Web address http://www.clf.org

Mailing address 62 Summer Street, Boston, MA 02110-1016

Phone number 617 350 0990

Fax number 617 350 4030

E-mail address members@clf.org

Council for a Livable World

This group was founded by the distinguished physicist Leo Szilard in 1962, when he felt that the danger of nuclear war was increasing.

Purpose CLW's main mission has been to combat the menace of nuclear war. Its principal activity to has been to channel financial campaign contributions to political candidates sympathetic to CLW's mission.

History The original purpose—to fight nuclear weapons—was broadened to cover also other weapons of mass destruction (chemical, biological). The Livable World Education Fund (tax deductible) was set up in 1980 to educate the public about the dangers of nuclear weapons and the arms race, and about peaceful alternatives. In the aftermath of the Gulf War, two new programs were created to work against arms sales to other governments and to build support for UN peacekeeping operations. The CLW has been a "scientist-led lobby of those who would contribute money to candidates committed to peace (Nichols 1974: 134)." It operates as a political action committee (PAC) that supports candidates for U. S. Senate. The affiliate committee PEACEPAC supports candidates for the House of Representatives. In the early period, the operations of CLW were "controlled by a self-perpetuating council of six scientists (ibid.: 139)," more recently by a Board.

Current status and activities CLW's projects in 1996 included an education campaign on arms control issues; monitoring and analysis of conventional arms sales and their impact; a public education campaign to build support for U.N. peace organizations; and efforts to educate the public and policy makers "about a military budget consistent with the real demands of US national security."

References

Lanouette, W. 1992. *Genius in the Shadows*. Scribner.

Nichols, D. 1974. The associational interest groups of American science. In *Scientists and Public Affairs*, ed. A. Teich. MIT Press.

Web address http://www.clw.org/pub/clw/welcome.html

Mailing address 110 Maryland Avenue, NE, Suite 409, Washington, DC 20002

Phone number 202 543 4100

E-mail address clw@clw.org

Council for Responsible Genetics

The scientists and other concerned citizens who started this organization in 1983 in Cambridge, Massachusetts, believed that decisions about genetic technologies were too important to be left in the hands of scientific "experts." The CRG was founded to help educate the citizenry about the social implications of technological innovations, and to empower that citizenry to have a voice in decision-making concerning technological developments and their implementation.

Purpose To serve as a national watchdog group focused on the biotechnology industry; to educate the public about the social and environmental implications of new genetic technologies, and to advocate for the socially responsible use of these new technologies.

History Much of CRG's early activities focused on organizing the scientific community to speak out against biological weapons and the militarization of biological research. These efforts culminated when CRG presented Congress with a petition signed by 2,000 scientists who pledged not to engage in research on biological weapons, and when CRG published a milestone book on this subject: *Preventing a Biological Arms Race*, ed. S. Wright (MIT Press, 1990). The CRG was the first organization to recognize, document, and define genetic discrimination, and remains the leading public-interest advocate for protections against this practice. The CRG developed model genetic discrimination legislation that has had an impact on political efforts in this area.

Current status and activities The CRG has five major program areas: (1) Maintaining an information clearinghouse for citizen activists, policy makers, and the media on social issues in genetics. (2) Preventing discrimination on the basis of predictive genetic tests. (3) Strengthening environmental regulation of commercial biotechnology. (4) Opposing the patenting of genetically engineered life forms and human genes. (5) Publishing GeneWATCH, the only national bulletin dedicated to both the social and environmental implications of biotechnology. The CRG has approximately 1,000 members and is headquartered in Cambridge, Massachusetts.

Web address http://www.gene-watch.org

Mailing address 5 Upland Road, Suite 3, Cambridge, MA 02140

Phone number 617 868 0870

Fax number 617 491 5344

E-mail address crg@gene-watch.org

Creation Research Society

This group was founded in 1963 in Michigan by ten scientists who had been unable to find publication outlets for their creationist research and saw the need for a journal that would accept this kind of work.

Purpose "Publication of a quarterly peer-reviewed journal; Conducting research to develop and test creation models; provision of research grants and facilities to creation scientists for approved research projects."

History The first issue of the *Creation Research Society Quarterly* was issued in July 1964. In 1996, a newsletter was added. From early on, CRS recognized the need for a research center, and around 1980 purchased land in Arizona on which the Van Andel Creation Research Center was built.

Current status and activities Memberships and subscriptions are at about 1,700 worldwide. CRS has 501(c)(3) not-for-profit tax-exempt status. In addition to the quarterly and the newsletter, CRS issues occasional publications and books.

Web address http://www.creationresearch.org/

Mailing address P.O. Box 8263, St. Joseph, MO 64508-8263

E-mail address contact@creationresearch.org

Education Development Center

In 1958, a group of MIT scientists joined forces with teachers and technical specialists and founded EDC to develop a new high school physics curriculum, called PSSC Physics. This curriculum focused on science as the product of experiment and theory, and was successfully introduced in schools across the US and abroad.

Purpose Using a broad and inclusive definition of education, EDC designs curricula and other educational tools and systems, and it also works to improve the conditions for learning and for the health of populations around the world.

History In the 1960s, EDC applied its approach to other subject areas and other countries. It created, for instance, an interdisciplinary social sciences program titled *Man: A Course of Study*, and devised the African Primary Science Program, which schools in eleven African countries used. Over time, EDC's range expanded to creating a variety of educational tools, from videotapes to computer software, and to addressing issues such as community health, nutrition, and democratization.

Current status and activities EDC, Inc., is a non-profit 501(c)(3) organization, with its main office in Newton, Massachusetts, and additional offices in Washington; New York; Carmel, California; and Newport, Rhode Island. It has a staff of more than 500 and carries out about 250 projects worldwide. "We conduct research and develop programs in such areas as early child development, K-12 education, health promotion, workforce preparation, learning technologies and institutional reform." It is characteristic of the approach of EDC to develop its programs in close collaboration and partnership with the people who will use them. On its Web site, EDC offers a large catalog of curricula, professional development materials and other publications.

Web address http://www.edc.org

Mailing address 55 Chapel Street, Newton, MA 02458-1060

Phone number 617 969 7100

Fax number 617 969 5979

E-mail address www@edc.org

Emergency Committee of Atomic Scientists

This group was founded in 1946 in Princeton, New Jersey, by Harold L. Oram (a New York fund raiser) and Albert Einstein. Einstein was the chairman.

Purpose Raising funds for FAS-related causes, including the *Bulletin of the Atomic Scientists* (Smith 1965).

History The great prestige of the committee members was garnered into initially quite successful fundraising efforts. In June 1947, an ECAS public statement about the uselessness of the United Nations Atomic Energy Commission and about the necessity of world government contradicted FAS positions and antagonized its leadership. In the wake of this disagreement, the activities of ECAS soon tapered off, and it officially folded in September of 1951.

References

Smith, A. 1965. *A Peril and a Hope*. University of Chicago Press.

Strickland, D. 1968. Scientists in Politics. Purdue University Studies.

Environmental Defense Fund (later Environmental Defense)

In 1966 a group of Long Islanders went to court to prevent the Suffolk County Mosquito Control Commission from spraying DDT on marshes. Arthur Cooley, Dennis Puleston, Charles Wurster, George Woodwell, and Robert Smolker presented scientific data to document the harmful impact of DDT on the osprey and other birds. Recognizing that their strategy of bringing scientific information to bear in the legal system held promise for wider application, individuals who had been involved in the successful court action formed EDF in the fall of 1967.

Purpose Protecting "the environmental rights of all people, including future generations," including "clean air, clean water, healthy, nourishing food, and a flourishing ecosystem." Sound science is to provide the rationale for evaluating and solving environmental problems. The current strategic plan of EDF identifies the following four major goals: stabilizing the earth's climate, safeguarding the oceans, protecting human health, and defending and restoring biodiversity.

History In its early years, EDF focused on the DDT problem and began a series of lawsuits across the US until a nationwide permanent ban on DDT was achieved in 1972. The public was particularly impressed with the finding, widely publicized by EDF, that DDT levels in mother's milk had risen to seven times the level permitted for milk sold in stores. In 1970, EDF took up the fight against airborne lead pollution, which resulted in an ultimately successful campaign to phase lead out of gasoline. Moreover, EDF focused on solving environmental problems in ways that also make economic sense. For instance, EDF convinced the California Public Utilities Commission that it would be more profitable to help consumers become more energy-efficient than to build additional power plants.

Current status and activities A nonprofit organization, EDF has its national headquarters in New York and a "Capital Office" in Washington, as well as a number of regional offices and project offices. The staff of 170 includes more than 75 scientists, economists, and attorneys. There are more than 300,000 members. It produces various publications and a bimonthly newsletter to the membership. It also uses the possibilities of the Internet, for instance by maintaining an electronic "Action Network" to influence national environmental policy, and by having a Chemical Scorecard Web site, on which people can find out about the pollution of their local area.

Web address http://www.edf.org

Mailing address 257 Park Avenue South, New York, NY 10010

Phone number 212 505 2100

Fax number 212 505 2375

E-mail address members@environmentaldefense.org

Environmental Law Institute

This institute was founded in 1969 in Washington.

Purpose To "advance environmental protection by improving law, policy, and management. . . . ELI researches pressing problems, educates professionals and citizens about the future of these issues, and convenes all sectors in forging effective solutions."

Current status and activities An independent legal research and education center, ELI has membership programs for individuals as well as institutions. Over its history, the Institute has trained more than 50,000 lawyers and other environmental professionals from all over the world. More than a third of ELI's work takes place outside the US. One of the current priorities is an expanded science and law program focused on risk. ELI's publications include the *Environmental Law Reporter* and the *Environmental Forum.*

Web address http://www.eli.org

Mailing address 1616 P St., NW, Suite 200, Washington, DC 20036

Phone number 202 939 3800

Fax number 202 939 3868

E-mail address law@eli.org

Environmental Literacy Council

Founded in 1996 in Washington, this was the successor to the Independent Commission on Environmental Education. The predecessor group had produced a report (Are We Building Environmental Literacy?) which concluded that environmental education often failed to introduce students to scientific and economic concepts essential for understanding environmental issues.

Purpose "To bring together scientists, economists, educators, and other experts to inform environmental studies."

Current status and activities ELC is a non-profit organization that helps locate and disseminate high-quality information about environmental issues. It works as a guide to environmental information and provides expert-review of educational materials. The ELC Web site includes an extensive list of links to pertinent sites.

Web address http://www.enviroliteracy.org

Mailing address 1730 K Street, NW, Suite 905, Washington, DC 20006-3868

Phone number 202 296 0390

Fax number 202 822 0991

E-mail address info@enviroliteracy.org

Ethical, Social and Legal Implications (Human Genome Project)

From its beginning, the Human Genome Project's leaders recognized the need to study its societal implications. "One of [James Watson's] first acts as director of the Human Genome Project was to announce that 3 percent (now 5 percent) of its funds would be set aside for studies of its ethical and legal consequences." (*New York Times*, April 7, 1998)

At its January 1989 meeting, the Program Advisory Committee on the Human Genome established a working group on ethics to develop a plan for the ELSI component of the overall project. In response to the working group's report, the National Human Genome Research Institute established the ELSI Branch (later renamed the ELSI Research Program) in 1990. The Office of Energy Research of the Department of Energy, which participates in the U.S. Human Genome Project, also reserved a portion of its funding for ELSI research.

Purpose ELSI was established to study the societal implications of the Human Genome Project.

History The four original priorities of ELSI were privacy and fair use of genetic information, clinical integration of genetic information into medicine, future research issues, and education. In the autumn of 1998, a year-long evaluation and review process of the ELSI program was concluded. The review resulted in five new goals (listed below).

Current status and activities ELSI is administered by the National Human Genome Research Institute (NHGRI), which belongs to the National Institutes of Health (NIH). NHGRI reserves 5 percent of its annual research budget for ELSI projects. By the end of 1998, the ELSI programs had funded almost $40 million worth of research. As stated on its website, its five priority areas for 1998-2003 are the following:

1. Examine the issues surrounding the completion of the human DNA sequence and the study of human genetic variation.
2. Examine issues raised by the integration of genetic technologies and information into health care and public health activities.
3. Examine issues raised by the integration of knowledge about genomics and gene—environment interactions into nonclinical settings.
4. Explore ways in which new genetic knowledge may interact with a variety of philosophical, theological, and ethical perspectives.
5. Explore how socioeconomic factors and concepts of race and ethnicity influence the use, understanding, and interpretation of genetic information, the utilization of genetic services, and the development of policy.

References

Rapp, R., D. Heath, and K. S. Taussig. 1998. Tracking the Human Genome Project. *Items* 52, no. 4: 88–91.

Web address http://www.nhgri.nih.gov/elsi/

Mailing address National Institutes of Health, Building 31, Room B2B07, 31 Center Drive, MSC 2033, Bethesda, MD 20892-2033

Phone number 301 402 4997

Fax number 301 402 1950
E-mail address elsi@nhgri.nih.gov

Ethics and Values in Science and Technology (now Societal Dimensions of Engineering, Science, and Technology) program (National Science Foundation)

In the early 1970s, a joint committee of NSF and the National Endowment for the Humanities was formed to study what NSF should do in the area of societal implications of science and technology. The committee recommended that a separate program and program director be created. Implementing this recommendation, NSF issued the first program announcement in fiscal year 1975, with the first awards in fiscal year 1976. The program was originally called "Ethical and Human Value Implications of Science and Technology," before the name changed to Ethics and Values in Science and Technology (EVIST). NSF had made awards in this subject area for several years prior to the establishment of the program.

Purpose "Developing and transmitting knowledge about ethical and value dimensions associated with the conduct and impacts of science, engineering, and technology."

History The EVIST program was renamed "Ethics and Values Studies" (EVS) and later merged with the "Research in Science and Technology" (RST) program to form the "Societal Dimensions of Engineering, Science, and Technology" (SDEST) program within the NSF Directorate for Social, Behavioral, and Economic Sciences.

Current status and activities Within the SDEST program, NSF makes approximately 40 new grants each year, with a budget of about $2.3 million.

Web addresses

http://www.cep.unt.edu/evs.html

http://www.nsf.gov/sbe/ses/sdest/start.htm

Mailing address 4201 Wilson Boulevard, Suite 995, Arlington, VA 22230

Phone number 703 306 1743

Fax number 703 306 0485

E-mail address rholland@nsf.gov

Federation of American Scientists

Manhattan Project scientists founded the Federation of Atomic Scientists as a civic organization that was licensed to lobby. The organizational meeting, which amalgamated a number of local scientists' groups, was held in Washington on October 31 and November 1, 1945. The *Bulletin of the Atomic Scientists*, also founded in late 1945, was the educational arm of the movement at that time. The Federation scientists opposed a War Department-sponsored bill that they feared would bring nuclear energy under a strong measure of military control (Nichols 1974: 132). They lobbied successfully for a civilian-run Atomic Energy Commission. However, farther-reaching ideas about an international control of atomic energy and about restricting this technology to non-military purposes remained unrealized. Largely the same scientists' groups that had formed the Federation of Atomic Scientists founded the Federation of American Scientists at a meeting held at George Washington University (in Washington, DC) on December 7 and 8, 1945. In 1946 the new organization absorbed the Federation of Atomic Scientists (Smith 1965).

Purpose The preamble to the 1945 FAS constitution named as the organization's primary purpose "to meet the increasingly apparent responsibility of scientists in promoting the welfare of mankind and the achievement of a stable world peace" (Coser 1965: 307; Smith 1965: 236–237). Currently, FAS describes its goal as "conduct[ing] analysis and advocacy on science, technology and public policy, including nuclear weapons, arms sales, biological hazards, secrecy, and space policy."

History Within the first two years of its existence, FAS membership reached 3,000 (Gottfried 1999: 43). During the 1950s and the 1960s, FAS worked through a combination of chapters in various cities and a small Washington office staffed mostly by volunteers. In 1970, it doubled the dues, hired a full-time director and began expanding its activities. At that time, it divested itself of a Los Angeles chapter (which became the Southern California Federation of Scientists) and of a Boston chapter.

Current status and activities For many years, Jeremy J. Stone was the driving force behind the organization; he served as its president from 1970 through 2000. Currently sponsored by about 60 Nobel Prize winners, its budget is under $1 million per year, and its staff of about a dozen works out of a set of townhouses on Capitol Hill in Washington. FAS tries to impact legislative arena through "direct lobbying, membership and grassroots work, and expert testimony at Congressional hearings." The recent program statement illustrates the wide scope of FAS interests: "Current war and peace issues range from nuclear war to ethnic conflict and from nuclear disarmament to arms sales; sustainable development issues include disease surveillance, climate modification, poverty, food security and environment. FAS also works on human rights of scientists and on reductions of secrecy." (*FAS Public Interest Report*, January-February 1998: 2, masthead) The primary publication of FAS is the *F.A.S. Public Interest Report.* The tax deductible arm of FAS is the FAS Fund, founded in 1972, which engages in research, analysis, and public education on a wide range of science, technology, and policy issues.

References

Coser, L. 1965. *Men of Ideas.* Free Press.

Gottfried, K. 1999. Physicists in politics. *Physics Today*, March: 42–48.

Lanouette, W. 1992. *Genius in the Shadows.* Scribner.

Lasby, C. 1966. Science and the military. In *Science and Society in the United States*, ed. D. Van Tassel and M. Hall. Dorsey.

Nichols, D. 1974. The associational interest groups of American science. In *Scientists and Public Affairs*, ed. A. Teich. MIT Press.

Smith, A. 1965. *A Peril and a Hope.* University of Chicago Press.

Strickland, D. 1968. Scientists in Politics. Purdue University Studies.

Wang, J. 1999. *American Science in an Age of Anxiety*. University of North Carolina Press.

Web address http://www.fas.org/about.htm

Mailing address 1717 K Street NW, Washington, DC 20036

Phone number 202 546 3300

Fax number 202 675 1010

E-mail address fas@fas.org

Forum on Physics and Society (American Physical Society)

Organized at the 1972 APS San Francisco meeting, as the first Forum of APS, "the FPS was born in the tumultuous 1960's and 70's. The issues of that era—the Vietnam War, the debate about the Anti-Ballistic Missile system, the energy crisis, the start of the environmental movement, the civil/human rights revolution—impelled that generation of physicists to consider their professional responsibilities."

Purpose FPS was founded "to address issues related to the interface of physics and society as a whole." Within the APS, the FPS provides a forum for examining topics, such as arms control, nuclear power, the efficient use of energy, and the environment.

History At first, the APS leadership eyed FPS with suspicion, because traditionally APS dealt only with professional issues. The APS Council appointed a senior APS member to attend the Forum Executive Committee meetings so as to make sure that the FPS did not overstep the APS's customary limits of concern. Over the years, however, the FPS gained the respect of its parent organization. Early leaders of FPS included Earl Callen (the founding chair), Martin Perl, Mike Casper, and Brian Schwartz.

Current status and activities FPS has about 4,500 members, which is roughly 11 percent of the total APS membership. One of the most important activities of FPS is to organize topical sessions at APS meetings. It also offers short courses to provide more in-depth background information on important issues. Two annual APS awards, the Burton Forum and the Szilard Awards, are sponsored by FPS. The Forum produces a newsletter titled *Physics and Society*, and also sponsors studies in its areas of interest. These studies and topical short courses are often published as books; for example: *The Energy Sourcebook*, ed. R. Howes and A. Fainberg (AIP 1991); *Global Warming*, ed. B. Levi et al. (AIP 1992).

Reference Hafemeister, D. History of the Forum on Physics and Society [on Web site].

Web address http://physics.wm.edu/~sher/phys_soc.html

Mailing address One Physics Ellipse, College Park, MD 20740

Phone number 301 209 3200

Fax number 301 209 0865

E-mail address peterz@erols.com

General Education in Engineering

In 1973, representatives of the engineering departments of 25 British universities and polytechnics met at Birmingham to found this group.

Purpose "To develop the full potential of engineering, not only as a professional study, but also as an exciting, worthwhile, and useful general education."

History GEE's activities included the sponsoring, discussion, and dissemination of study materials concerned, for example, with entrepreneurialism in engineering and with the issues surrounding the professional status and responsibilities of engineers, as well as the encouragement of socio-technical projects as part of engineering degree courses. The project was managed by a Coordinator, David Brancher of the University of Aston in Birmingham, assisted by a Council. Within GEE, the Socio-Technical Projects Programme was coordinated by Frank Swenson of the University of Stirling. The project was funded by the Nuffield Foundation for three years. During that time, it held a number of workshops and open meetings, produced two sets of project outlines, two project registers, and made a number of small grants and awards to both departments and students to assist and reward their efforts. The various documents produced by GEE were deposited with the librarian at the Department of Electrical Engineering, Imperial College, London.

George C. Marshall Institute

The institute was set up in 1984. The impetus for its founding was the perception that, in our civilization, which more and more depends on science and technology, scientific appraisals are often politicized and misused by interest groups.

Purpose The Institute's goal is "to preserve the integrity of science and promote scientific literacy". To that end, it seeks to provide "policymakers with rigorous, clearly written and unbiased technical analyses on a range of public policy issues."

History A major focus of the Institute's work has been the issue of global warming. The scientists associated with the Institute have tended to be skeptical about the accuracy and scientific validity of forecasts of human-induced global warming, and critical of what they consider hasty policy reactions. Other areas of interest have included the future of the space program and defense issues. Among the Institute's recent major projects was an *Independent Commission on Environmental Education*, which conducted a content evaluation of environmental education resources and released a report of its findings ("Are We Building Environmental Literacy?") (see Web site).

Current status and activities The George C. Marshall Institute is a nonprofit 501(c)(3) corporation. Its activities include briefings to the press, publication programs (including a newsletter), speaking tours, and public forums, such as the "Washington Roundtable on Science and Public Policy."

Web address http://www.marshall.org/

Mailing address 1730 K Street, NW, Suite 905, Washington, DC 20006-3868

Phone number 202 296 9655

Fax number 202 296 9714

E-mail address info@marshall.org

Hastings Center

The Hastings Center was founded in 1969 by the philosopher Daniel Callahan and the physician Willard Gaylin.

Purpose To conduct research on "the broad range of problems and issues posed to our society by developments on medicine and the life-sciences—from test-tube babies to genetic engineering, organ transplantation, end-of-life decisions, health care rationing, and more."

History The Hastings Center has played an important role in the debate about rDNA research (Krimsky 1982: 22), the legal definition of death, the use of life-sustaining medical treatment, priority setting in health care delivery, health policy reform, and the teaching of ethics in medical schools and other professional settings.

Current status and activities The Hastings Center has recently broadened its scope to include biotechnology and institutional changes in the health care system. A few examples of current projects are studies of decisions near the end of life, dying with dementia, ethical issues in managed health care, and values in biotechnology. Research projects are carried out by research groups composed of Center staff, Center Fellows (distinguished experts who work closely with the center) and other individuals with relevant expertise. The Center also conducts workshops, conferences, and educational programs, such as Visiting Scholar and Internship programs.

Reference Krimsky, S. 1982. *Genetic Alchemy*. MIT Press.

Web address http://www.thehastingscenter.org

Mailing address Route 9-D, Garrison, NY 10524

Phone number 914 424 4040

Fax number 914 424 4545

E-mail address mail@thehastingscenter.org

Hudson Institute

In 1961 the physicist and futurist Herman Kahn founded the Hudson Institute in collaboration with Max Singer and Oscar Ruebhausen.

Purpose "Forecasts trends and develops solutions for governments, businesses, and the public."

History Until his death in 1983, Kahn was the driving force of the Institute. Since then the Institute has continued to conduct research along the lines of Kahn's interest in the future, while also addressing significant challenges of the present. In 1962, the Hudson Institute relocated from New York City to Croton-on-Hudson; and in 1984, the Institute's headquarters moved to Indianapolis. Over the years, the Hudson Institute produced numerous reports on important issues of public interest. The following are a few examples. In 1962, *Thinking about the Unthinkable*, an analysis of nuclear warfare by Herman Kahn, appeared. In 1968, the Hudson Institute published *Can We Win in Vietnam?* The research staff was divided between those favoring and those opposing the war. The *Why ABM?* study in 1969 expressed support for the deployment of ABM missiles. In 1987, a major work titled *Workforce 2000* forecast trends in the American workforce. Its sequel *Workforce 2020* was issued in 1997. In 1994, the Hudson Institute cooperated with Wisconsin in reorganizing this state's welfare system.

Current status and activities Headquartered in Indianapolis, the Hudson Institute also has an office in Washington, and six other offices worldwide. It is a not-for-profit 501(c)(3) organization, with more than 70 researchers and employees. Grants and donations from US and international companies and individual supporters are its major sources of funding. Its publications include a quarterly magazine titled *American Outlook*. The Hudson Institute is equally focused on domestic and international issues. It describes itself as a "nonpartisan futurist organization that is known for its healthy skepticism of conventional wisdom and its guarded optimism about solving tomorrow's problems today."

Web address http://www.hudson.org

Mailing address 5395 Emerson Way, Indianapolis, IN 46226

Phone number 317 545 1000

Fax number 317 545 9639

E-mail address info@hudson.org

Institute for Science and International Security

Founded in 1993, ISIS is "a non-profit, non-partisan institution dedicated to informing the public about science and policy issues affecting international security. Its efforts focus on stopping the spread of nuclear weapons, bringing about greater transparency of nuclear activities worldwide, and achieving deep reductions in nuclear arsenals. ISIS's projects integrate technical, scientific, and policy research in order to build a sound foundation for a wide variety of efforts to reduce the threat posed by nuclear weapons to US and international security."

History ISIS has produced internationally recognized technical assessments of countries' efforts to acquire nuclear weapons. It has often been at the forefront of efforts to solve complex national and international security problems. Both in the US and abroad, it has tried to unite government officials, independent experts, scientists, and the public in efforts to find credible strategies to solve US, regional, and global security problems. In early 1999, following the onset of military strikes in Yugoslavia, ISIS worked with US government officials and the media to ensure that a nuclear research facility near Belgrade would not be bombed by NATO warplanes. ISIS also encouraged the International Atomic Energy Agency (IAEA) to resume inspections of the facility, which has enough weapon-grade uranium for two nuclear weapons. ISIS initiated efforts in mid-1998 to regularly update information about civil plutonium stocks worldwide, and in 1997, ISIS published the most complete history of Taiwan's nuclear weapons efforts. ISIS has also been active in assessing Iraq's nuclear weapons program. David Albright served as the first (and so far only) nongovernmental inspector on an IAEA Action Team mission in that country. More recently, ISIS has collaborated with a former senior Iraqi scientist who defected from Iraq, and also with German technicians who provided Iraq with sensitive information and items related, in particular, to centrifuge uranium enrichment technology. ISIS has had a long-standing interest in North Korea's plutonium production. It recently concluded a multi-year assessment of North Korea's indigenous nuclear program, its efforts to deceive the IAEA about its program, and the need to verify this program, as called for under the Agreed Framework. A book-length report of this assessment, titled *Solving the North Korean Nuclear Puzzle*, was released in the autumn of 2000. ISIS has also been in the forefront of using commercial satellite imagery to detect nuclear proliferation activities. In particular, ISIS used this technology to pinpoint the first Pakistani test site. It also revealed a heavy water plant near the Pakistani Khushab nuclear reactor.

Current status and activities Currently, ISIS's two principal projects are the Nuclear Nonproliferation Project, which examines how nuclear weapons have spread and how these processes can be halted and reversed, and the Nuclear Weapons Production Project, which analyzes the policies of the US nuclear weapons complex, encourages nuclear transparency, and seeks the broad application of international controls on nuclear explosive materials. ISIS hopes to learn from regions where nuclear weapons have been eliminated and apply these lessons to current regions of tension, such as the Middle East and the Korean Peninsula. ISIS also provides information and advice to governmental bodies that seek to establish greater government openness and public accountability. ISIS publishes fact sheets and reports such as *Challenges of Fissile Material Control* and *Plutonium Watch*, and maintains a database of more than 5,500 policy makers, experts, scientists, journalists, and the interested public. ISIS staff members present research at conferences and workshops throughout the world, and provide technical information to the US government and many print and broadcast media organizations. ISIS publishes many of its findings in journals such as the *Bulletin of the Atomic Scientists*, and through its own publications by the ISIS Press. ISIS regularly employs student interns for three-to-four month periods, and occasionally hosts fellows with special needs for access to ISIS resources or other resources unique to Washington. ISIS is also a member organization of the Herbert Scoville Jr. Peace Fellowship Program.

Web address www.isis-online.org/

Mailing address 236 Massachusetts Ave. NE, Suite 500, Washington, DC 20002

Phone number 202 547-3633

Fax number 202 547-3634

E-mail address isis@isis-online.org

Institute on Religion in an Age of Science

The two parent groups of IRAS were the American Association for the Advancement of Science's Committee on Science and Values and an interfaith coalition. Members of both groups wished for a more constructive relationship between science and religion. In 1954, the scientists accepted the religious group's invitation for a seven-day conference on "Religion in an Age of Science" on Star Island (near Portsmouth, New Hampshire). Encouraged by this meeting, members of the two groups decided to form IRAS as a continuing organizational framework for the positive exchange between science and religion.

Purpose "Working for a dynamic and positive relationship between religion and science."

History In 1964, IRAS spawned the *Center for Advanced Study in Religion and Science* (CASIRAS).

Current status and activities IRAS's main activity is an annual week-long conference on Star Island on topics that are relevant to the interconnection of religion and science. It co-publishes the quarterly journal *Zygon*, and it puts out a newsletter. It also maintains electronic discussion forums, and sponsors other meetings, conferences, and discussion groups.

Web address http://www.iras.org

Intergovernmental Panel on Climate Change

In 1988, recognizing the risk of a potential global climate change, the World Meteorological Organization (WMO) and the United Nations Environment Programme (UNEP) established the IPCC.

Purpose "To assess the scientific, technical and socio-economic information relevant for the understanding of the risk of human-induced climate change." IPCC does not carry out new research, nor does it monitor data, but it bases its assessments mainly on the published and peer-reviewed expert literature.

History IPCC completed its First Assessment Report in 1990. This played a key role in negotiating the UN Framework Convention on Climate Change, which entered into force in 1994. The Second Assessment Report, issued in 1995, was a major input for the Kyoto Protocol in 1997.

Current status and activities IPCC is open to all members of the UNEP and WMO. It has three Working Groups and a Task Force. Working Group I assesses the scientific aspects of the climate system and of climate change. Working Group II addresses the vulnerability of socio-economic and natural systems to climate change, the consequences of climate change, and options for adapting to it. Working Group III assesses options for limiting greenhouse gas emissions and for other ways of mitigating climate change. The Task Force oversees the National Greenhouse Gas Inventories Programme. Apart from the Assessment Reports, IPCC also prepares special reports and technical papers.

Web address http://www.ipcc.ch

Mailing address 7bis Avenue de la Paix, C.P. 2300, 1211 Geneva 2, Switzerland

Phone number +41 22 730 8208

Fax number +41 22 730 8025

E-mail address ipcc_sec@gateway.wmo.ch

International Council for Science

Founded in 1919 as the International Research Council, this group provides an international forum for learned academies and scientific societies. It works toward breaking the barriers of specialization between the scientific disciplines, and it promotes the application of scientific advances for the benefit of humanity. It also champions freedom in the conduct of science.

History In 1931, the International Research Council was reorganized into the International Council of Scientific Unions. The US affiliates were the National Academy of Sciences and the National Research Council. In 1946, the ICSU committee on science and social relations published a survey of scientific associations about their views on problems arising from research. ICSU later became affiliated with the United Nations Educational, Scientific and Cultural Organization (UNESCO). In 1998, the name was changed from International Council of Scientific Unions to International Council for Science, while the acronym ICSU was retained.

Current status and activities The ICSU comprises 98 multi-disciplinary associations (e.g., national academies of science), 26 international single-discipline scientific unions, and 28 scientific associates. ICSU organizes hundreds of scientific meetings and congresses each year around the world, as well as major international, interdisciplinary research programs, such as the International Geosphere-Biosphere Programme: A Study of Global Change (IGBP). ICSU maintains close working relations with intergovernmental and non-governmental organizations, especially with UNESCO. It publishes a wide range of newsletters, handbooks, and journals.

References

Smith, A. 1965. *A Peril and a Hope*. University of Chicago Press.

ICSU Annual Report 1999.

Web address http://www.icsu.org

Mailing address 51 Bd. de Montmorency, 75016 Paris, France

Phone number +33 145250329

Fax number +33 142889431

E-mail address secretariat@icsu.org

International Dark-Sky Association

Purpose To raise awareness about the problem of light pollution of the night skies, and to work toward solutions. IDA promotes quality nighttime lighting which both decreases light pollution and improves energy efficiency. IDA emphasizes that its goals have a public-interest aspect. Light pollution and other sources of interference not only hamper astronomers, but also make life worse for people in general. They prevent them from enjoying the beauty of the night sky. Moreover, the development of more efficient lighting that reduces light pollution would also mean significant energy savings.

History Since its inception in 1988, IDA has held annual meetings in Tucson. A number of IDA sections have come into existence and have also organized local and regional meetings.

Current status and activities IDA is a tax-exempt nonprofit 501(c)(3) organization for educational and scientific purposes. All the officers and board members are volunteers. In 2000, IDA had more than 3,300 members from the US and 69 other countries. Among the members are about 200 groups or organizations. IDA issues newsletters and information sheets. Its other activities include a Speakers' Bureau, Good Lighting Awards, and the publicizing of good lighting design. In addition to combating light pollution and promoting quality outdoor lighting, IDA is also building awareness of radio frequency interference, space debris, and other obstacles that impede the view and study of the skies.

Web address http://www.darksky.org/~ida/

Mailing address 3225 N. First Avenue, Tucson, AZ 85719

Phone number 520 293 3198

Fax number 520 293 3192

E-mail address ida@darksky.org

International Institute for Applied Systems Analysis

In 1966, in the context of new initiatives on East-West bridge building, President Lyndon Johnson approved a proposal to initiate negotiations for a multi-national research institution. The goal was the establishment of a non-governmental forum where Eastern and Western scientists could investigate techniques for the efficient management of large programs and enterprises common to all modern societies. Protracted negotiations ultimately led to the founding of IIASA on October 4, 1972. Representatives of twelve nations signed the charter establishing IIASA "to initiate and support collaborative and individual research in relation to problems of modern societies arising from scientific and technological developments." Founding members were the USA, the USSR, Bulgaria, Canada, Czechoslovakia, France, the Federal Republic of Germany (FRG), the German Democratic Republic (GDR), Italy, Japan, Poland, and the United Kingdom. IIASA membership is vested in a professional, non-governmental institution from each member nation, which acts as the National Member Organization (NMO), although funding generally comes from the respective government.

Purpose IIASA provides science-based policy insight to government and industry leaders who make decisions about complex issues, such as energy use, food security, population, health, and research priorities. The Institute was founded on the belief that joint research among national institutions speeds improvements in essential methods of analysis and investigation, and promotes the spread of shared scientific understanding, thus facilitating joint action among members.

History The Institute's original projects focused on methodological and applied research in systems analysis, operations research, and management techniques in such fields as water resources, energy systems, urban and regional systems, and the design and management of large organizations. In the 1970s and 1980s, IIASA attracted respected scientists from around the world, whose work at IIASA achieved both academic and practical success. Examples include the following: In 1975 a new research field, adaptive Ecosystem Policy and Management, was founded at

IIASA and was used for forest management policy. IIASA models in energy assessment and agricultural and forestry systems analysis have affected the thinking of policy makers worldwide since the 1970s. In 1989, the Regional Air Pollution Information and Simulation (RAINS) model was adopted as the basis for the negotiation of a sulfur dioxide emissions protocol by the UN Economic Commission for Europe (UN-ECE), and IIASA's energy models have been used by the World Energy Congress to develop scenarios out to the year 2050. IIASA's summer fellowship program for advanced graduate students, the Young Scientists Summer Program (YSSP), has grown from its initial ten participants in 1977 to more than 50 in the 1990s. IIASA's political situation has likewise evolved in the 25 years since its founding. New NMOs joined IIASA from Austria (1973), Hungary (1974), Sweden (1976), Finland (1976) and the Netherlands (1977). In 1982, the Reagan administration, concerned at reports that Eastern Bloc intelligence agents had infiltrated IIASA, withdrew US government support from IIASA. At that time, the American Academy of Arts and Sciences replaced the National Academy of Sciences as the US National Member Organization, and the American dues were raised through private contributions. Britain withdrew entirely in 1982. In 1989, the US government conducted a review of the IIASA connection, and renewed American government support. IIASA's focus shifted with the end of the Cold War, and its 1991 strategic document *Agenda for the Third Decade* reflected a new emphasis on global environmental, economic, and technological changes.

Current status and activities The Institute is currently undergoing another redefinition and renewal of its focus. The future work will build on the massive databases and complex modeling that characterize many of IIASA's existing projects, and that constitute the core of the Institute's strength. A few examples will suggest the ongoing and emerging directions of the Institute's work. The Environmentally Compatible Energy Strategies project had two books published by Cambridge University Press in late 1998, one presenting 50–100-year worldwide energy projections, the second a historical look at technology and development. The RAINS model remains a major tool for European negotiations on limiting transboundary air pollution. A new Social Security Reform project has begun to model scenarios for the global impacts of the aging of societies, and the Risk, Modeling and Policy Project is developing models to guide the design of financial instruments for insuring developing nations against massive losses associated with natural disasters. The YSSP program continues to garner support from member nations and IIASA researchers, and participants consistently give glowing reviews of their IIASA experience. IIASA publishes a quarterly magazine called *Options* in addition to the scholarly articles and reports produced by the projects. The magazine and many of the reports are available at the IIASA Web site. IIASA is overseen by a Council with one member from each National Member Organization (NMO). The US NMO is the American Academy of Arts and Sciences, which fulfills its responsibilities through the US Committee for IIASA.

References

Charter of the International Institute for Applied Systems Analysis 1972.

25th Anniversary (1998) Pamphlet on IIASA history (available from IIASA).

IIASA 1999 Research Plan. Summary of planned research for 1999.

McDonald, A. 1998. The International Institute for Applied Systems Analysis. Talk given at New York Academy of Sciences conference on Scientific Cooperation and Conflict Resolution.

Raiffa, H. 1992. The Founding of IIASA. Talk given at IIASA, September 23 [on Web site].

Web address http://www.iiasa.ac.at/

Mailing address A-2361 Laxenburg, Austria

Phone number +43 2236 807 0

Fax number +43 2236 71313

E-mail address inf@iassa.ac.at

US National Member Organization US Committee for IIASA (c/o American Academy of Arts and Sciences, 136 Irving St., Cambridge, MA 02138; phone 617 576 5000; fax 617 576 5050; e-mail mcollins@amacad.org)

International Network of Engineers and Scientists for Global Responsibility

Founded 1991, this is "an association of member organizations and individual members committed to a responsible use of science and technology."

Purpose "Promotes international communication and cooperation to achieve these goals [of responsible use of science and technology]. The efforts of INES focus on disarmament, conversion, non-proliferation, ethics and sustainable development."

History In 1995, to mark the fiftieth anniversary of the first nuclear explosion, INES appealed to engineers and scientists to sign a pledge in which they promised to uphold social responsibility and ethical standards in their work.

Current status and activities Recognized by the UN as an official NGO, INES is currently a world-wide association of more than 90 organizations and 250 individual members. It is led by a Council and an Executive Committee. Among the US organizational affiliates are the American Engineers for Social Responsibility (AESR). There are sponsoring members who contribute money, and members who participate in INES's projects. The current activities center upon conversion (from military to civilian production), non-proliferation, ethics, international an sustainable development, and the "Buddhist perception of nature." INES has a central network office in Dortmund, Germany, and largely self-reliant network groups.

Web address http://www.mindspring.com/~us016262/ines.html

Mailing address Gutenbergstrasse 31, 44139 Dortmund, Germany

Phone number +49 231 575202

Fax number +49 231 575210

E-mail address INES_NAT@t-online.de

International Peace Research Institute, Oslo

Purpose "Studying the causes and consequences of war and peace."

History Founded in 1959, this organization—now widely known as PRIO—became a fully independent institute in 1966. Three years later, a six-week summer course in peace research was inaugurated. (This summer program still exists.)

Current status and activities PRIO is an independent, international research institute, with a staff of 60, of whom 35 are researchers and guest researchers. In addition, there are 8 graduate students from the University of Oslo. In a 1998 reorganization, four Strategic Institute Programs were established: Conditions of War and Peace; Foreign and Security Policies; Ethics, Norms and Identities; Conflict Resolution and Peacebuilding. PRIO issues a wide range of publications, among them two international journals, *Journal of Peace Research* and *Security Dialogue.*

Web address http://www.prio.no/default.asp

Mailing address Fuglehauggata 11, 0260 Oslo, Norway

Phone number +47 22547700

Fax number +47 22547701

E-mail address info@prio.no

International Physicians for the Prevention of Nuclear War

Bernard Lown of the Harvard School of Public Health and Yevgeny Chazov of the Cardiology Research Center in Moscow founded this organization. The two internationally renowned physicians had known each other since 1960 and had been research collaborators. Lown was also among the founders of Physicians for Social Responsibility. In 1979, Lown proposed to Chazov to organize an international movement of physicians to combat the nuclear threat. Together with four other physicians (two Americans and two Soviets), the two men set up IPPNW the next year in Geneva.

Purpose IPPNW was instituted to mobilize the medical community against the threat of nuclear warfare. It currently defines itself as "a global federation of national physicians' organizations dedicated to safeguarding health through the prevention of war."

History In 1984, IPPNW received the Peace Education Prize of UNESCO, and, in the following year, the Nobel Peace Prize. By 1985, IPPNW had more than 135,000 members in 41 countries (including 28,000 in the US and 60,000 in the USSR.). A major activity of IPPNW has been to convene international congresses about the medical aspects of nuclear war. It has also advocated a freeze on the development and deployment of nuclear weapons, a no-first-use pledge by the nuclear powers, and the diversion of funds from nuclear weapons to social and health problems. During the Cold War, IPPNW maintained a position of neutrality, refraining from supporting or attacking the positions of any involved government.

Current status and activities IPNNW is run by an international board of directors (with three Co-Presidents), and an Executive Committee. It has more than 80 national affiliates, with its headquarters located in Cambridge, Massachusetts. IPPNW publishes *Vital Signs*, the *Global Health Watch* series, and numerous research studies. In addition to its continuing efforts towards the abolition of nuclear weapons, the organization currently campaigns for a ban of land mines and conducts a Peace Through Health program.

Reference T. Wasson, ed. 1987. *Nobel Prize Winners*. H. W. Wilson.

Web address http://www2.healthnet.org/IPPNW/

Mailing address 727 Massachusetts Avenue, Cambridge, MA 02139

Phone number 617 868 5050

Fax number 617 868 2560

E-mail address ippnwbos@igc.apc.org

International Science and Technology Center

To support the nonproliferation of high-technology weapons of mass destruction after the collapse of the Soviet Union, the European Community, Japan, the US, and Russia created this intergovernmental organization in 1992. Operations began in 1994.

Purpose Under the motto "Nonproliferation through science cooperation," ISTC provides weapons scientists and technical team members in the former Soviet Union with opportunities to direct their scientific talents to peaceful scientific projects.

History Since 1992, the number of member nations has expanded. Norway and the Republic of Korea joined, as well as five CIS countries (Armenia, Belarus, Georgia, Kazakstan, and Kyrgyzstan). Through March 2000, ISTC programs have funded 940 projects valued at $267 million. More than 30,000 individuals have received grant payments.

Current status and activities At the core of ITSC is its Science Project Program, which supports research in civilian fields. The Partner Program facilitates collaboration between scientific institutions of the participating countries. A wide range of supporting programs and activities, from business management training and patenting support to travel grants, is available. ITSC also monitors the funded projects to ensure that they comply with the program goals.

Web address http://www.istc.ru

Mailing address Luganskaya Ulitsa 9, 115516 Moscow, Russia

Phone number +7 501 797 6010

Fax number +7 501 797 6047

E-mail address istcinfo@istc.ru

Loka Institute

The perceived lack of attention to the environmental, social, and political effects of scientific and technological progress, as well as the perceived lack of ordinary citizens' involvement in the scientific and technological decision-making processes prompted

Richard E. Sclove to found this institute in 1987. The name Loka is derived from the Sanskrit word 'lokasamgraha', which means unity of the world, interconnectedness of society, and duty to perform action for the benefit of the world.

Purpose "To make science and technology more responsive to social and environmental concerns by expanding opportunities for grass-roots, public-interest group, everyday citizen, and worker involvement in vital facets of science and technology decision making."

History In 1997, Loka organized a pilot Citizens' Panel on telecommunications and the future of democracy. This panel was fashioned after the "consensus conference" model, pioneered in Denmark and introduced to the US by Loka's president Richard Sclove. More panels of this kind are planned.

Current status and activities The Loka Institute is a nonprofit [501(c)(3)] research and advocacy organization, with a staff of almost ten, plus interns. It makes extensive use of the Internet: There are Loka Alerts, FASTnet (Federation of Activists on Science and Technology Network), the Community Research Network List, and Pol-sci-tech (a global interactive discussion list). Loka also briefs the government, gives public presentations and media interviews, organizes conferences, and serves as an informal information clearing house.

Web address http://www.loka.org

Mailing address P.O. Box 355, Amherst, MA 01004-0355

Phone number 413 559 5860

Fax number 413 559 5811

E-mail address Loka@Loka.org

Médecins Sans Frontières

Médecins Sans Frontières (Doctors Without Borders) was founded in 1971 in a merger of two groups of physicians, both based in France, who intended to provide efficient emergency assistance wherever wars and other disasters occurred. The first group had worked in Biafra from 1968 through 1970 for the Red Cross; the second had assisted the victims of a tidal wave that hit eastern Pakistan (the future Bangladesh) in 1970. Both groups were dissatisfied with the kind of international aid available at the time. They thought it included too little medical assistance and, in addition, was hampered by administrative and diplomatic obstacles.

Purpose The organization "delivers medical relief to populations in danger due to war, civil strife, epidemics or natural disasters." It observes strict neutrality and impartiality. Its medical assistance is provided by volunteers.

History The organization remained very small during the early 1970s. The established international organizations tended to regard the MSF volunteers as amateurs or "medical hippies" (Brauman 1993). A 1976 mission in Lebanon provided some publicity, and, in 1978, the organization started to expand considerably. Key factors behind this growth were a sharp rise in the refugee population worldwide and a glut in the labor market for physicians in France. Its expansion caused the organization to become more professional. Internal tensions developed that led to the exodus of most of the organization's founders. Modern fund-raising techniques were introduced in 1982. Learning from its missions, MSF became expert in improving the logistical and technical aspects of emergency relief. Although MSF never suspends neutrality in terms of the treatment it provides, it did, in some cases, denounce the atrocities and human rights violations its volunteers witnessed. In the 1980s, chapters of Médecins Sans Frontières were set up in a number of countries. The US branch, started in 1990, has attracted 160,000 individual donors since its founding. MSF was awarded the 1999 Nobel Peace Prize.

Current status and activities The American affiliate of Médecins Sans Frontières has offices in New York and Los Angeles. It is led by a board of directors and an Advisory Board. The international organization has a budget of about $250 million and sends roughly 2,500 volunteers to some 80 countries annually. Although care for refugees remains an important focus for Médecins Sans Frontières, the organization increasingly faces the challenge of new heath care crises, such as the reemergence of once-controlled diseases, the emergence of new epidemics (especially AIDS), and the lack of access to health care for the world's poor. Developing countries have traditionally been the focus of Médecins Sans Frontières activities, but with civil war in ex-Yugoslavia and the collapse of communism, Eastern Europe and the former Soviet Union have also become major target areas. In the U. S., Doctors Without Borders raises funds, fosters public awareness, and recruits American volunteers. In 1997, the American affiliate raised $8.4 million in financial support, and nearly 100 American volunteers participated in field projects.

Reference Brauman, R. 1993. The Médecins Sans Frontières experience. [on Web site]

Web address http://www.dwb.org/

Mailing address 6 East 39th Street, 8th floor, New York, NY 10016

Phone number 212 679 6800

Fax number 212 679 7016

E-mail address dwb@newyork.msf.org

Medical Committee for Human Rights

Doctors and nurses active in the civil rights movement formed MCHR in 1964.

Purpose The original mission was "to provide medical care during the civil rights demonstrations and marches (Nichols 1974: 152)." MCHR had the goal of using medical skills in the service of progressive causes, and, ultimately, saw itself as a participant in an attack on the capitalist economic system.

History By 1971, MCHR had 24 local chapters, which were largely autonomous (Nichols 1974: 152). In the early 1970s, the membership was about 6,000. By early 1971, MCHR ran about 125 free health clinics, which were mostly set up by community groups). MCHR also promulgated radical critiques of American society and profit-based heath care (Krimsky 1982: 19). MCHR was known for demonstrations at American Medical Association (AMA) conferences (Nichols 1974: 155). At its April 1971 convention, MCHR decided to focus on a nationally coordinated campaign to supplement local activism. MCHR published the newspaper *Health Rights News* and the magazine *The Body Politic.*

References

Krimsky, S. 1982. *Genetic Alchemy*. MIT Press.

Nichols, D. 1974. The associational interest groups of American science. In *Scientists and Public Affairs*, ed. A. Teich. MIT Press.

National Action Council for Minorities in Engineering

Founded in 1974, this is a not-for-profit organization that receives support from corporations, government agencies, universities, alumni, and individuals. It advances its mission through partnerships with industry, government, and the education community.

Purpose "To increase the representation of successful African Americans, Latinos and American Indians in the most vital of the nation's professions, engineering."

Current status and activities "NACME conducts research, analyzes and advances public policies, develops and operates precollege, university and workforce programs." Among the organization's publications are the *NACME Research Letters*. Conferences and the Internet are other means of dissemination. NACME engages in training activities, such as diversity and mentor training, and organizes a number of scholarship and fellowship programs. Since 1980, nearly 10 percent of all minority engineering students graduated from college with NACME scholarships

Web address http://www.nacme.org

Mailing address The Empire State Building, 350 Fifth Avenue, Suite 2212, New York, NY 10118-2299

Phone number 212 279 2626

Fax number 212 629 5178

National Center for Science Education

When bills promoting creationism in schools were introduced in several state legislatures in the late 1970s and the early 1980s, citizens who wished to oppose the creationism movement formed local "Committees of Correspondence." In 1981, several of these committees founded NSCE, which was incorporated in 1983.

Purpose "To provide a central information and resource clearinghouse, helping to coordinate the efforts of people working at state and local levels to preserve the integrity of science education."

Current status and activities NCSE is a nonprofit, tax-exempt membership organization. In addition to defending the theory of evolution against attacks, it also works to increase public understanding of evolution and science. The "Human Evolution Education Network" brings together K-12 teachers and scientists who want to help teach evolution in the classroom. NCSE has an on-line bookstore offering relevant publications, publishes a bi-monthly journal titled *Reports of the National Center for Science Education*, pamphlets, and instructional materials for the teaching of evolution. It disseminates information about creationism and the theory of evolution in the mass media, and supports local groups in opposing creationism. Members include not only scientists, but people from all walks of life.

Web address http://www.NatCenSciEd.org/

Mailing address P.O. Box 9477, Berkeley, CA 94709-0477

Phone number 510 526 1674

Fax number 510 526 1675

E-mail address ncse@NatCenSciEd.org

National Conference of Lawyers and Scientists

This group was established by the American Association for the Advancement of Science and the American Bar Association in 1974.

Purpose "To promote a better understanding of science among lawyers and judges and of the legal system among scientists; To improve communications between lawyers and judges on the one hand and scientists and engineers on the other; To monitor and examine emerging public policy issues of concern to both lawyers/judges and scientists/engineers; To examine such issues cooperatively and, where appropriate, to recommend policy alternatives to their respective organizations and others relating to such matters; To sponsor joint symposia, programs and studies; and To identify and collaborate with groups from other nations interested in exploring similar subjects."

Current status and activities The NCLS has 14 members, half appointed by AAAS and half by the ABA Science and Technology Section, who meet four times a year. The NCLS sponsors symposia, conferences, workshops, and projects, several of which have resulted in publications. Recent topics of interest include scientific misconduct, enhancing the availability of scientific information in the courts, and the legal and ethical aspects of computer networks, of electronic science publishing, and of advances in genetics.

Web address http://www.aaas.org/spp/dspp/sfrl/committ/ncls.htm

Mailing address:

1200 New York Ave. NW, Washington, DC 20005

Phone number 202 326 6600

Fax number 202 289 4950

E-mail address mfrankel@aaas.org

National Society of Black Engineers

Appalled by the 80 percent dropout rate of black freshmen in engineering at Purdue University, two undergraduates, Edward Barnette and Fred Cooper, started the Black Society of Engineers (BSE) at that institution in 1971. This student organization was intended to improve the recruitment and retention of black engineering students. As the on-campus activities of BSE proved successful, it reached out to other schools to form a national association. In April 1975, the NSBE was founded during a meeting hosted by BSE, which 48 students from 32 schools attended. The leaders in creating the national organization were Anthony Harris, Brian Harris, Stanley R. Kirtley, John W. Logan, Edward A. Coleman, and George A. Smith. In 1976, NSBE was incorporated as a non-profit organization in Texas.

Purpose "To increase the number of culturally responsible Black engineers who excel academically, succeed professionally and positively impact the community"; to increase the numbers of minority students in all engineering disciplines.

Current status and activities NSBE is headquartered in Alexandria, Virginia, and has a national staff of about 30. It is mainly a student association, but also includes an alumni extension. There are more than 13,000 members in more than 280 student chapters and 75 alumni (technical professionals) chapters. NSBE organizes twelve regional conferences and an annual convention. Other activities include tutorial programs, group study sessions, high school/ junior high outreach programs, technical seminars and workshops, a national communications network (NSBENET), two national magazines (*NSBE Magazine* and *NSBE Bridge*), an internal and a professional newsletter, resume books, career fairs, and awards.

Web address http://www.nsbe.org

Mailing address 1454 Duke St., Alexandria, VA 22314

Phone number 703 549 2207

Fax number 703 683 5312

E-mail address member@nsbe.org

National Women's Health Resource Center

The recognition that women lacked easy access to reliable and comprehensive health information and resources prompted the establishment of NWHRC. It was incorporated in the summer of 1997 as part of the Columbia Hospital for Women in Washington. The organization was intended to give consumers one central place to turn to with their health questions.

Purpose NWHRC considers itself "the national clearinghouse for information and resources about women's health." Its primary goal is "to educate healthcare consumers and empower them to make intelligent decisions" by providing easy-to-understand and easy-to-reach information and services.

Current status and activities NWHRC, Inc., is a not-for-profit organization with corporate headquarters in New Brunswick, New Jersey, and offices in Washington. It has a Women's Health Advisory Council representing a wide range of health specialties, and a Corporate Advisory Board to establish links with private industry. NWHRC publishes the *National Women's Health Report*.

Web address http://www.healthywomen.org

Mailing address 120 Albany Street, Suite 820, New Brunswick, NJ 08901

Phone number 877 986 9472

Fax number 732 249 4671

E-mail address info@healthywomen.org

Natural Resources Defense Council

The planned construction of Storm King (a pump storage power plant on the Hudson River) and the Alaskan oil pipeline were among the issues that motivated the founding of NRDC in 1970. John Adams, a lawyer, has been its CEO from the beginning.

Purpose "To safeguard the Earth, its people, its plants and animals, and the natural systems on which all life depends." In protecting the environment, the NRDC also seeks "to break down the pattern of disproportionate environmental burdens borne by people of color and others who face social or economic inequities."

History In its early years, NRDC focused on drafting major environmental legislation, such as the Clean Air Act and the Clean Water Act. As more and more legislation was being passed, NRDC increasingly functioned as a legal "watchdog" that made sure that the environmental laws were properly applied. Starting out as an organization of lawyers, NRDC began adding scientists to its staff in the early 1970s. It also was a non-member organization at the outset, basically operating like a law firm, but soon started to recruit members to diversify and stabilize its financing (rather than relying solely on foundation grants). Over its history, NRDC got involved in many of the major environmental issues. To give a recent example, NRDC settled a lawsuit it had brought against the EPA in early 2000. The settlement required the EPA to close some loopholes in the Clean Water Act. In March 2000, NRDC was instrumental in stopping the development of a salt works in Baja California—the last pristine calving ground for Pacific gray whales.

Current status and activities A not-for-profit, tax-exempt membership organization, NRDC is headquartered in New York and has regional offices in Washington, San Francisco, and Los Angeles. It has a staff of 175 and more than 400,000 members nationwide. In 1999, its budget was about $30 million. NRDC's four major strategies are scientific research, advocacy and activism, lobbying, and litigation. It produces a variety of publications, among them the quarterly *Amicus Journal*, which is also presented on the Web site.

Web address http://www.nrdc.org

Mailing address 40 West 20th Street, New York, NY 10011

Phone number 212 727 2700

Fax number 212 727 1773

E-mail address nrdcinfo@nrdc.org

New University Conference

Founded in March 1968, NUC was a nationwide campus organization of college and university faculty and graduate students, with the goal of advancing a wide spectrum of radical policies and politics.

History NUC's existence coincided with the height of political activism in the late 1960s and early 1970s. On the basis of leftist convictions, NUC developed position papers on issues ranging from campus politics, to domestic policies, to the Vietnam war. Its emphasis on feminism was notable. NUC had obvious affinities and connections with the more radical wing of the scientist activists. NUC was committed to the ideals of participatory democracy. There were unresolved tensions between members who emphasized the autonomy of the local chapters, and the national leadership who tried to build a national organization. By 1971, NUC had a paid membership of about 2,000, with chapters on 60 campuses; by June 1972, it had only 300 members and was bankrupt. The last NUC convention decided to disband. NUC left a legacy of campus-based child care centers, graduate assistant unions, and radical professional caucuses.

References

Krimsky, S. 1982. *Genetic Alchemy*. MIT Press.

Pincus, F., and Ehrlich, H. 1988. The New University Conference: A study of former members. *Critical Sociology* 15, no. 2: 145–147.

NIH Black Scientists Association

This is an autonomous organization, recognized by the National Institutes of Health, whose membership includes more than 70 scientists, physicians, and other professionals at NIH. Its activities center on information dissemination, including a seminar series titled "Science Working For Us," on networking, career support, and advocacy regarding issues of importance to underrepresented minorities at NIH and beyond. NIH BSA also maintains an e-mail network called BSCINET (Black SCIentist interNETwork).

Purpose "To get to know each other as people and as professionals, to promote our individual and collective professional advancement, and to advocate various health and scientific issues of importance to underrepresented minority communities in general and to the Black community in particular."

Web address http://www.nih.gov/science/blacksci/

Mailing address P.O. Box 38, Clarksburg, MD 20871

Phone number 301 496 3027

E-mail address th112c@nih.gov

Office on Public Understanding of Science (National Academy of Sciences)

This office was created in response to three perceived needs: a need to increase scientists' willingness and ability to present their work to wider publics; a need to assess, aid, and improve science reporting in the media; and a need to increase appreciation for science in the general public. Its establishment was approved by the Academy Council of the NAS on June 14, 1996.

Purpose "To foster the mutual responsibility of scientists and the media to communicate to the public, with accuracy and balance, the nature of science and its processes as well as its results."

Current status and activities OPUS manages the "Beyond Discovery" project consisting of a series of published case studies that trace the origins of recent technological and medical breakthroughs, and thereby reveal the crucial role played by basic science in these advances. OPUS also co-sponsors a "Distinguished Leaders in the Life Sciences" lecture series, and managed a competition for television and web producers that recognized innovative science TV and web programming.

Web addresses

http://www4.nas.edu/nas/opus.nsf

http://www.BeyondDiscovery.org

Mailing address 2101 Constitution Avenue, NW, Washington, DC 20418

Phone number 202 334 1575

Fax number 202 334 1690

E-mail address opus@nas.edu

Peace Action (formerly SANE/Freeze)

In 1957 the publication of Albert Schweitzer's "Call to Conscience" in *Saturday Review* brought the dangers of nuclear radiation to public attention. *Saturday Review* editor Norman Cousins and Clarence Pickett of the American Friends Service Committee called a meeting at the New York apartment of the poet Lenore Marshall, and the National Committee for a Sane Nuclear Policy (SANE) was founded. It attracted a number of prominent members and became an effective national voice for nuclear disarmament.

Purpose The mission of Peace Action is to "work through national and grassroots citizens' action to promote global nuclear disarmament, reduce unnecessary military spending, and end the international arms trade."

History In the 1960s and the 1970s, SANE expanded its mission, joined the anti-Vietnam War movement, and formed alliances with organized labor. In the early 1980s, the Nuclear Weapons Freeze Campaign was initiated by Randall Forsberg. Freeze was a grassroots-based confederation of groups, with offices in St. Louis and Washington, and was active in mobilizing citizens against the nuclear arms race and the threat of nuclear war. In 1987, SANE and Freeze merged. The name of the new organization was initially SANE/Freeze, but changed to Peace Action in 1993.

Current status and activities Peace Action currently has more than 60,000 members, 27 state affiliates, and 125 chapters. PA has many scientists as members, although it is not, per se, a scientists' group (Nichols 1974: 134). Among its recent and current concerns are cutting the military budget; nuclear abolition; a halt to weapons trafficking; a campaign against the sanctions on Iraq; and the Peace Voter '98 Campaign (which includes Voter Guides, paid advertisements, and "vigilant bird-dogging"). The national office in Washington, houses an Organizing Department that fosters activism across the nation. PA also publishes an annual voting record of federal legislators. PA's tax-deductible sister organization is the Peace Action Education Fund (PAEF). PAEF educates the public and policy makers about peace and disarmament issues and puts out fact sheets and briefing papers. PAEF has an international office in New York to monitor peace and disarmament issues in the United Nations. A quarterly newsletter is published jointly by PA and PAEF. In addition, the quarterly mailings of PA's Grassroots Network against the Arms Trade

(GNAT) include in-depth information on the global arms trade and on strategies to challenge it. The Peace Empowerment Program provides monthly action alerts on priority policy issues.

Reference D. Nichols. 1974. The associational interest groups of American science. In *Scientists and Public Affairs*, ed. A. Teich. MIT Press.

Web address http://www.webcom.com/peaceact/

Mailing address 1819 H Street, Suite 420, Washington, DC 20006-3606

Phone number 202 862 9740

Fax number 202 862 9762

E-mail address pamembers@igc.apc.org

Physicians for a National Health Program

Health care professionals associated with Harvard University founded PNHP in 1987. A "single issue organization advocating a universal, comprehensive single-payer national health care program," it works for a "national health care reform based on principles of social justice and medical need." In 1989, PNHP published the first major single-payer proposal for health care reform in the *New England Journal of Medicine*. This and related proposals on health care reform have sparked discussion and media coverage.

Current status and activities PNHP has a membership of 9,000 physicians and other health care workers. There are more than two dozen local chapters. PNHP activities include presentations at grand rounds, conferences, and other professional forums, outreach to the media, coalition work with other public-interest groups, the development of health policy proposals, and two annual meetings. PNHP maintains a Speakers' Bureau on health care reform, and provides access to resource materials, among them slide sets. Members receive a quarterly newsletter.

Web address http://www.pnhp.org

Mailing address 332 S. Michigan, Suite 500, Chicago, IL 60604

Phone number 312 554 0382

Fax number 312 554 0383

E-mail address pnhp@aol.com

Physicians for Human Rights

Founded in late 1986, PHR established an office in Somerville, Massachusetts, with Jonathan Fine as the first executive director. Throughout its history, PHR has intervened in numerous countries where human rights violations have occurred. In 1988, the membership reached 1,000, PHR attracted grants from several major donors, and the staff doubled from two to four. By 1993, the membership grew to 4,000. Together with Human Rights Watch, PHR co-authored *Landmines: A Deadly Legacy* (1993). In 1997, PHR was among the recipients of the Nobel Peace Prize for its work on the steering committee of the International Campaign to Ban Landmines.

Purpose "Uses the knowledge and skills of the medical and forensic sciences to investigate and prevent violations of international human rights and humanitarian law." PHR has a policy of impartiality and opposes human rights violations regardless of the involved governments or groups.

Current status and activities PHR is headquartered in Boston and also maintains a Washington office. With a staff of about 20, PHR focuses on chemical weapons, forensic science/war crimes investigations, landmines, medical ethics and human rights, medical neutrality, prisons, public health effects of conflicts, refugees/immigration, and torture. It conducts educational and training projects for health professionals, legal professionals, and human rights advocates on the application of medical and forensic skills in the investigation of human rights violations. PHR issues wide range of publications, among them a PHR newsletter called *The Record*, and maintains a Health and Human Rights database of pertinent Web sites. It also organizes a Speaker's Bureau on human rights issues, and circulates Action Alerts, mostly to urge members to write letters on behalf of persecuted medical professionals.

Web address http://www.phrusa.org

Mailing address 100 Boylston Street, Suite 702, Boston, MA 02116

Phone number 617 695 0041

Fax number 617 695 0307

E-mail address phrusa@phrusa.org

Physicians for Social Responsibility

This group—founded in 1961, chiefly by Harvard faculty in the medical fields (Nichols 1974: 147–148)—was originally modeled after the Society for Social Responsibility in Science (SSRS). In the early 1970s, it had about 900 members. PSR has been an affiliate of International Physicians for the Prevention of Nuclear War, which was founded in 1980 and which received the Nobel Peace Prize for 1985. Harvard School of Public Health Professor Bernard Lown was a driving force in both organizations. In the mid 1980s, PSR moved its national office to Washington. From 1984 to 1992, PSR was part of the Professional Coalition for Nuclear Arms Control, a lobbying coalition focusing on strategic nuclear weapons. In 1993, PSR broadened its mission in response to global threats to health and security in the post-cold-war period. In 1996, PSR helped to secure its long-sought goal of a Comprehensive Nuclear Test Ban Treaty (which, however, failed to be ratified by the US Senate), and, in 1997, joined efforts in Kyoto, Japan, to achieve a Global Climate Change Convention.

Purpose "To create a world free of nuclear weapons, global environmental pollution, and gun violence. The active conscience of American medicine, PSR uses its members' expertise and professional leadership, influence within the medical community and strong links to policy makers to address this century's greatest threats to human welfare and survival."

Current status and activities PSR has 15,000 members nationwide, 15 professional staff members at its national headquarters in Washington, led by an executive director, about 40 active chapters, and some 10 full- and part-time staff members in local offices. The board of directors consists of 32 doctors and two medical students and has an annually elected Board President. PSR currently focuses on issues of national security, violence prevention, and the environment and health. PSR provides an electronic service ("ACT!") on its Web site that alerts members and the public to current events of PSR interest and urges them to educate politicians. PSR has a monthly newsletter, *PSR Reports*, for its members, and a bulletin, *PSR Monitor*, which provides news on public policy issues. It is associated with International Physicians for the Prevention of Nuclear War (IPPNW) and the International Society of Doctors for the Environment (ISDE).

References

Nichols, D. 1974. The associational interest groups of American science. In *Scientists and Public Affairs*, ed. A. Teich. MIT Press.

Cortright, D. 1993. *Peace Works*. Westview.

Web address http://www.psr.org/

Mailing address 1101 14th St., NW, Suite 700, Washington, DC 20005

Phone number 202 898 0150

Fax number 202 898 0172

Pugwash Conferences

The initial impulse for the Pugwash Conferences came from the Russell-Einstein manifesto, which Einstein signed two days before his death in 1955. The manifesto included a call upon the international scientific community to assemble and discuss the threat to civilization brought on by nuclear weapons. Bertrand Russell issued the invitations to such a conference in 1957 and became chairman of the Continuing Committee. The distinguished biophysicist Joseph Rotblat took a leading role in this initiative (and was to remain at the helm of the Pugwash Conferences for many years). American industrialist and philanthropist Cyrus Eaton, who since 1955 had brought together various groups of scholars and educators at Pugwash, Nova Scotia (his birthplace), provided the financing and hospitality for the conference. The first conference gave rise to a series of successor conferences, which have retained the name of the first venue (Pugwash) although they have been held in various venues around the world.

Purpose "Pugwash is a union of scientists who are concerned about the relations between science and society. Its purpose is not to promote the interest of scientists, nor to fight for the status of science, to discuss salaries or funds for research. Instead we are alarmed by the fact that science, which for so long was thought to confer nothing but blessings to mankind, today also displays a different aspect. It can be exploited—and is exploited—for destruction and repression. . . . The most serious problem facing mankind today is how science can be used, not for warfare, but for the welfare of the human race" (H. Alfvén, in foreword to Rotblat 1972). "The purpose of the Pugwash Conferences is to bring together, from around the world, influential scholars and public figures concerned with reducing the danger of armed conflict and seeking cooperative solutions for global problems." Prominent scientists and senior policy figures attend Pugwash meetings as individuals, not as representatives of their governments, for off-the-record discussions of major scientific and policy issues. Then these persons, again in their individual capacity, communicate the results and insights of Pugwash meetings to their national policy makers and representatives of international organizations. Because of the participants' elevated stature, insights from the Conferences often penetrate into official policy-making.

History Three committees were formed at the first conference, dealing with (1) hazards arising from the use of atomic energy in peace and war; (2) control of nuclear weapons; and (3) social responsibility of scientists (Rotblat 1972: 5). The Pugwash agenda has been heavily slanted toward security and disarmament issues—primarily focusing on the nuclear threat, but also including biological and chemical warfare (e.g., 5th Conference, 1959). Among other topics of interest have been international scientific cooperation (e.g., 7th Conference, 1961), and developmental aid (15th, 1965) (21st, 1971) (Rotblat 1972: vii–viii). Pugwash has also reached out into the environmental sphere (e.g., 46th Conference, 1996, with a working group on energy/climate interaction). In addition to its conferences, Pugwash has in recent years undertaken some more sustained research projects and published their results. Since the late 1950s, the American Academy of Arts and Sciences has been the institutional home of US Pugwash. Pugwash is credited with helping to lay the groundwork for various arms control treaties, including the 1972 Anti-Ballistic Missile Treaty, the SALT accords, and the Chemical Weapons Convention. It was awarded the Nobel Peace Prize for 1995 (one half of the prize went to the Pugwash Conferences, the other half to Joseph Rotblat).

Current status and activities By the end of 1998, Pugwash had convened a total of 246 conferences, workshops, and special symposia, in which about 3,600 individuals from 108 countries participated (many of those attended more than one event so that the total attendance count exceeds 10,000). Around the world, there are more than 30 national Pugwash groups, with an additional 20 Student/Young Pugwash groups. The American student group, Student Pugwash USA, which was founded in 1979 at the University of California at San Diego, now has 45 chapters and annually about 1,500 to 2,000 participants in its activities. Funding for the Pugwash Conferences and related activities has come from various sources, among them the American Academy of Arts and Sciences and the MacArthur Foundation. Pugwash publishes a variety of reports and monographs, including the *Pugwash Newsletter*, the *Proceedings of the Annual Pugwash Conference*, the *Annals of Pugwash*, and special monographs on such topics as nuclear weapons and proliferation, chemical and biological weapons, and military research and development. The central element of Pugwash continues to be that prominent scientists and senior policy makers attend Pugwash meetings as individuals for off-the-record discussions of major scientific and policy issues. The Pugwash organizational structure consists of a President, Secretary-General, Council, and Executive Committee, with small permanent offices in London, Geneva, and Rome. Many functions of the Pugwash secretariat were recently relocated to the American Academy of Arts and Sciences in Cambridge, Massachusetts. At the Academy, actions with respect to Pugwash fall under the Committee on International Security Studies (CISS).

References

Holdren, J. 1995. Nobel Prize acceptance speech. Web site.

Moore, M. 1997. Forty years of Pugwash. *Bulletin of Atomic Scientists* 53, November-December: 40–45.

Rotblat, J. 1967. *Pugwash*. Czechoslovak Academy of Sciences.

Rotblat, J. 1972. *Scientists in the Quest for Peace*. MIT Press.

Rotblat, J. 1995. Nobel Prize Acceptance Speech. Web site.

Rotblat, J., ed. 1997. *World Citizenship*. Macmillan.

Zinberg, D. 1997. Pugwash: Quiet force for change. *Times Higher Education Supplement*, September 12.

Web address http://www.pugwash.org

Mailing addresses

c/o American Academy of Arts and Sciences, 136 Irving St., Cambridge, MA 02138; phone 617 576 5022; fax 617 576 5050; e-mail pugwash@amacad.org

Flat A Museum Mansions, 63A Great Russell Street, London WC1B 3BJ, England; phone +44 171 4056661; fax +44 171 8315651; e-mail pugwash@qmw.ac.uk

Red Crate Collective

In the 1970s, the Red Crate Collective was a small group of left-wing citizen-scientists centered in the Boston area. Its purpose was to provide humanitarian help, at a modest scale, to developing socialist countries. For instance, it shipped used textbooks to Cuba and Vietnam. The books were packed in red crates.

Reference Krimsky, S. 1982. *Genetic Alchemy*. MIT Press.

Science and Environmental Policy Project

This group was founded in 1992 in Arlington, Virginia, by the atmospheric physicist S. Fred Singer, on the premise that "sound, credible science must form the basis for health and environmental decisions that affect millions of people and cost tens of billions of dollars every year."

Purpose "To clarify the diverse problems facing the planet and, where necessary, arrive at effective, cost-conscious solutions." SEPP opposes the influence of the environmentalist activist movement, charging that its lobbying has often created economically wasteful policies on the basis of inadequate or inconclusive scientific evidence. In general, SEPP rejects what it considers "regulatory excess" by the government, especially the Environmental Protection Agency, on issues such as ozone depletion, air pollution, and acid rain. It favors a market approach to natural resource management over government regulations. A core issue for SEPP is climate change and global warming. Here the group argues that scientific evidence does not support the claims of a human-made global warming ("greenhouse effect").

History To counteract the perception that the scientific community had reached a consensus about the existence of human-induced global warming, SEPP circulated a statement, called the Leipzig Declaration (referring to a November 1995 conference in Leipzig, Germany), that expressed concern about hasty policy action based on inadequate science. The declaration attracted more than 100 signatures from climate scientists. More recently, a petition against the climate accord reached in Kyoto, Japan, drew 17,000 signatories.

Current status and activities The main focus of SEPP remains the global warming debate. SEPP is also in favor of a vigorous space program that would include Mars exploration and the protection of Earth against asteroid impact. SEPP is a non-profit, non-partisan educational group. It has assembled a Board of Science Advisors. SEPP publishes books and research reports in the field of environmental science, and organizes seminars and conferences. Its expertise is sought after by policy makers and the news media. SEPP has openings for graduate and undergraduate interns throughout the year.

Reference Singer, S. 1997. *Hot Talk, Cold Science*. Independent Institute.

Web address http://www.sepp.org/

Mailing address 4084 University Drive, Suite 101, Fairfax, VA 22030

Phone number 703 934 6932

Fax number 703 352 7535

E-mail address comments@sepp.org

Science and Human Rights Program (American Association for the Advancement of Science)

Founded in 1976, the Science and Human Rights Program has five objectives: (1) protecting the human rights of scientists; (2) advancing scientific methods and skills for documenting and preventing human rights violations; (3) developing scientific methodologies for monitoring the implementation of human rights; (4) fostering greater understanding of and support for human rights among scientists; (5) conducting research on human rights issues.

History In 1992, the Program began publishing a directory of persecuted scientists, engineers, and health professionals, documenting its casework. In 1993, the AAAS Human Rights Action Network (AAASHRAN) was started. Through e-mail, it informs AAAS members and other subscribers of cases and developments that deserve special attention, and coordinates the recipients' letter-writing efforts on behalf of persecuted colleagues. In 1993, the Program began collaborating with the International Center for Human Rights Investigations in Guatemala. This work culminated in the 1999 report State Violence in Guatemala, 1960 to 1996. Over its history, the Program assisted a number of NGO and governmental institutions in human-rights-related matters.

Current status and activities The Science and Human Rights Program is a unit of the AAAS Directorate for Science and Policy Programs. Besides support through the regular AAAS budget, the Program has received funding from a great number of foundations and from individual donors. The Program sponsors research and meetings within the range of its objectives, and issues a newsletter titled *The Report on Science and Human Rights*. On its Web site, the Program maintains an on-line Directory of Human Rights Resources, the largest Internet listing of human rights organizations with a World Wide Web presence.

Sources See Standing Committee on Scientific Freedom and Responsibility.

Web address http://shr.aaas.org

Mailing address 1200 New York Ave. NW, Washington, DC 20005

Phone number 202 326 6600

Fax number 202 289 4950

E-mail address shrp@aaas.org

Science in a Social Context

A group of academics from ten British institutions (Aston, Edinburgh, Leeds, Leicester, London, Manchester, Middlesex, Stirling, Surrey, Sussex) founded "SISCON" at a meeting held at the University of Leeds in 1972. The group, comprising scientists as well as sociologists, historians, and philosophers of science, were seeking an alternative to the conventional technical teaching of science at the undergraduate level, and wished to see more emphasis on the social aspects. Some courses along these lines were already in place at a few British universities, notably Manchester and Edinburgh, and SISCON wanted this approach to be adopted more widely.

Purpose To change the undergraduate teaching of science from the purely technical to one that combines the techniques of science with the consideration of its social aspects.

History In the period 1973–76, SISCON was funded by the Nuffield Foundation. It was controlled by a council (with one member from each founding institution) and a coordinator (Bill Williams). As interest in the project grew, members of other institutions were co-opted onto the Council. The initial effort concentrated on producing teaching/learning texts. Research Fellows were employed to write up short modules of the kind that had proved successful in Manchester, Edinburgh and elsewhere. In this way, more than thirty of such modules were produced and distributed widely both in the UK and abroad. It was also intended to support experts from one institution to give a series of lectures in another, but this happened only once. There was an annual conference and summer school. The Leverhulme Trust funded a second three-year period, Michael Gibbons of Manchester University became Coordinator, and there was less emphasis on the production of modules. The annual conferences/summer schools continued. Two summer schools were held in Amsterdam, in collaboration with a Dutch group and with Dutch funding. Eleven of the projects modules were revised and published by Butterworth. SISCON effectively disbanded in 1980. There is still some demand for the modules. The original Coordinator, Bill Williams, still holds substantial quantities of some of the initial trial modules and of some of the Butterworth versions. There were also corresponding efforts at the secondary school level. The *Siscon-in-Schools* Project was an offshoot of SISCON, whereas two other groups, *Science and Society* and *Science and Technology in Society (SATIS)*, independently pursued similar interests. There is also a SISCON equivalent in engineering, called *General Education in Engineering* (GEE) [see there].

Mailing address c/o Dr. W. F. Williams, Centre for Joint Honours in Science, University of Leeds, Leeds LS2 9JT, England

Phone number +44 113 275 1618

Fax number +44 113 233 2689

E-mail address cbs7wfw@leeds.ac.uk

Scientific Freedom, Responsibility and Law Program (American Association for the Advancement of Science)

This program came into existence in 1990. Its predecessor was the AAAS Office of Scientific Freedom and Responsibility.

Purpose The Scientific Freedom, Responsibility and Law Program is "committed to upholding high ethical standards for science and engineering and to examining the impediments to and opportunities for achieving those standards; to improving understanding of the relationship between the law and science; to monitoring ethical, legal, and social issues related to science and technology and bringing them to the attention of scientists, engineers, policymakers, and the public; to recognizing the actions of persons whose behavior has fostered scientific freedom and responsibility; and to facilitating the exchange of ideas and information related to the Program's mission among all interested parties."

History Continuing the work of its predecessor, the Program has undertaken numerous projects, among them examinations of the ethical and legal implications of genetic testing, and of computer and network use and abuse, as well as an investigation into the effects of political and economic changes in Eastern Europe on the scientific freedom and responsibility of scientists in that area.

Current status and activities The Scientific Freedom, Responsibility and Law Program is a unit of the AAAS Directorate for Science and Policy Programs. It works in close collaboration with the AAAS Committee on Scientific Freedom and Responsibility and the National Conference of Lawyers and Scientists (a joint committee of AAAS and the American Bar Association's Science and Technology Section). The Program also coordinates the Professional Society Ethics Group, which provides a forum for representatives from professional societies to discuss issues of professional ethics, and publishes the quarterly newsletter, *Professional Ethics Report.*

Sources See entry for Standing Committee on Scientific Freedom and Responsibility.

Web address http://www.aaas.org/spp/dspp/sfrl/sfrl.htm

Mailing address 1200 New York Ave. NW, Washington, DC 20005

Phone number 202 326 6600

Fax number 202 289 4950

E-mail address mfrankel@aaas.org

Scientists and Engineers for Social and Political Action / Science for the People

This organization was formed by physicists attending the 1969 American Physical Society meeting in New York, originally under the name "Scientists for Social and Political Action." The initiative came from faculty members and graduate students in the San Francisco Bay area, who opposed the anti-ballistic missile (ABM) program and the Vietnam War, and wanted the APS to speak out on these issues. After the APS membership had in June 1968 voted down a constitutional amendment (proposed by Charles Schwartz) that would have allowed APS officially to take stances on political concerns, the activist physicists formed a new organization outside the APS. Among the leaders were Charles Schwartz (Berkeley), Martin Perl (Stanford), Marc Ross (Michigan), and Michael Goldhaber (Rockefeller University) (Moore 1996: 1609–1610).

Purpose Although initially SESPA was not radical in conception, it almost immediately became radicalized (Nichols 1974: 155–156), and the group's purpose then was rooted in a sweeping sociological analysis of science. Science, SESPA members argued, was anything but a politically neutral institution. Its ultimate function was to prop up capitalism; and it served this function mainly by supporting and legitimizing government projects and policies. Because the root problem was thus identified as pervasive and systemic, SESPA also called for a sweeping response. "We scientists are workers. Our only hope in preventing further misuse of science is to join with all other workers to bring about a radical change in the thinking, goals, and economic structure of this country." (cited in ibid.: 156) Over time, the critique of science increased in complexity, including the role science was said to play in supporting imperialism, racism, sexism, and poverty, and in polluting the environment. SESPA's mission was to demystify science and teach science to "the people." It also promoted uses of science that would benefit "the people" and not the government, and opposed what it considered the misuses of science by the government, and especially by the military. The group was self-consciously radical, in its purpose as well as in its methods.

History The geographic center of the organization shifted to the Boston area. Important events in the early formation of SESPA were the 1969 and 1970 American Association for the Advancement of Science meetings in Boston and Chicago, respectively, and the 1971 SESPA meeting in Cambridge. SESPA became notorious for disrupting official AAAS events. The organization was highly decentralized and loose-knit; among the local groups, Boston predominated. SESPA provided free or low-cost mathematics and science courses and published a magazine called *Science for the People*. During the early years (1969–1972), the group's activities focused on ending the Vietnam War and on cutting the ties between science and the military. For example, SESPA in New York pressured Columbia University to get rid of the Pupin Lab, funded almost completely by the Department of Defense. In Philadelphia, SESPA members were involved in defusing bombs at bomb factories. The "Science for Vietnam Project" of the Chicago group sent books, scientific supplies, and medical supplies to scientists and doctors in Vietnam. This group also had a policy of non-cooperation with government scientists. Teaching subgroups in various cities wrote pamphlets and gave presentations on how to teach "progressive" science, and other subgroups provided technical support to local radical organizations and groups. Local groups were often associated with international radical science groups, as well as many New Left organizations. After 1972, SftP formed two loosely associated sub-units, *Science for the People* (which produced the magazine) and the *Science Resource Service* (which engaged in activism and educational programs). As the antiwar movement wound down, the issues of women in science shifted into the foreground. The SftP became more and more inward looking; internal controversies about its identity and mission heated up. In the 1980s there were attempts to re-energize the organization by revitalizing the ailing magazine, creating a national organization, and achieving financial stability. In 1989, the IRS began an audit that drove the organization into bankruptcy and dissolution.

References

Moore, K. 1993. Doing Good While Doing Science: The Origins and Consequences of Public Interest Science Organizations in America, 1945–1990. Dissertation, University of Arizona.

Nichols, D. 1974. The associational interest groups of American science. In *Scientists and Public Affairs*, ed. A. Teich. MIT Press.

Scientists for Global Responsibility

This group was formed in London in 1992 in a merger of Scientists Against Nuclear Arms, Electronics and Computing for Peace, and Psychologists for Peace. The merger took place at a meeting held in the University of London Union. To some extent, it was an act of desperation on the part of SANA, which was in danger of folding.

Purpose SGR promotes "the socially responsible and ethical use of science and technology." It lists the following as its major activities: "promote a widespread debate among 'experts' and public about the proper role of science and technology; increase public awareness of important new scientific work; raise issues concerning democracy, decision-making, public information, science and ethics; question the much-vaunted concept of scientific 'objectivity' and develop a new vision for science and scientists appropriate for a socially responsible, democratic society."

History The issues surrounding nuclear arms and nuclear war were at the core of SANA's activities. Especially with the merger into SGR, the organization's focus has widened considerably to include environmental concerns and, in general, the ethical and policy questions concerning science.

Current status and activities SGR is the UK affiliate of the International Network of Engineers and Scientists for Global Responsibility (INES), and has about 1,000 members. According to a 1995 membership survey, SGR members were most likely to have a background in physics; computing was a close second. The most sizable area of employment was higher education; the second largest category consisted of retirees. Among the wide range of members' interests, the environment and nuclear weapons were mentioned most frequently. SGR organizes conferences and disseminates a Newsletter and other publications. A recent publicity-gaining activity was riding the "Climate Train to Kyoto," when the UN Climate Convention was held there, in an effort to lobby for climate protection. A recent focus of SGR's work is an ethics initiative that raises awareness about ethical issues in science and offers a list of mentors and consultants in this area. SGR initiates the founding, at universities, of committees on ethics in science. A project on the measurement and the impacts of electromagnetic fields is in preparation.

Reference Forum InformatikerInnen für Frieden und gesellschaftliche Verantwortung, SGR-Portrait (http://hyperg.uni-paderborn.de/fiff/organisationen/sgr)

Web address http://www.gn.apc.org/sgr/

Mailing address Unit 3 Down House, The Business Village, Broomhill Road, London SW18 4JQ, England

Phone number +44 181 871 5175

E-mail address sgr@gn.apc.org

Scientists for Orlov and Shcharansky (later Scientists for Sakharov, Orlov, and Shcharansky)

The imprisonment in 1977 of Soviet physicist Yuri Orlov, the head of the Moscow-Helsinki Watch Group, outraged the American scientific community. In 1978, two physicists at the Lawrence Berkeley National Laboratory, Morris Pripstein and Denis Keefe, felt compelled to come to the support of Orlov and other imperiled colleagues behind the Iron Curtain, including Natan (Anatoly) Shcharansky. With the help of Berkeley Laboratory scientists, including Andrew Sessler and Owen Chamberlain, they founded Scientists for Orlov and Shcharansky, a movement that within 2 years of its small-scale beginnings swelled into a world-wide protest movement of scientists.

Purpose The goal of SOS was to secure the release of scientific dissidents from Soviet oppression.

History The "moratorium concept," developed by Pripstein to pressure the Soviet government, distinguished SOS from other efforts. The idea was to ask members of the scientific community to freeze their scientific exchanges with Soviet scientists. Within days of its announcement, the moratorium attracted the pledges of 500 scientists. After Shcharansky's trial for treason, 2,400 scientists joined the moratorium. In 1980, when Andrei Sakharov was exiled to Gorky, SOS added his name to its title. The number of moratorium signatories rose to more than 8,000 scientists from 44 countries. SOS is credited with a major role in the eventual liberation of the three most prominent dissidents, as well as of a number of lesser known ones. Orlov was released in 1986, and Shcharansky and Sakharov soon followed.

Reference Friedlander, M. 1998. Morris Pripstein wins Human Rights Award for effort to help Soviet dissidents. Berkeley Lab Science Articles Archive (http://www.lbl.gov/Science-Articles/Archive/pripstein-sakharov.html).

Scientists' Institute for Public Information

In the late 1950s and early 1960s, the St. Louis area experienced high levels of radioactive fallout, and concerned citizens demanded more information about this perceived hazard. In 1963 a group of scientists from Washington University formed SIPI as a coalition of 23 local fallout information groups.

Purpose SIPI's mission was to provide the public with unbiased scientific information. It held that making scientific facts available to the citizens and thus enabling them to make informed decisions is decisive in keeping democratic government viable. SIPI itself typically shied away from taking political stances. "With respect to the resultant value judgments [on the basis of scientific information], scientists have no greater or lesser competence than other informed citizens and ought not to arrogate such decisions to themselves." (Moore 1993: 217–218)

History By early 1964, the group had broadened its emphasis from the genetic and environmental effects of radiation to environmental issues in general. SIPI's membership was committed to the principle of volunteer participation, and opposed to centralization and professionalization. SIPI had a staff of only two; its leaders were Barry Commoner and Margaret Mead. Conflicts over the identity of the group arose at the end of the 1960s, centering on the issues of scientists vs. non-scientists, and of local groups vs. national organization. Moreover, SIPI's periodical *Environment* (earlier called *Nuclear Information* and *Scientist and Citizen*) was a financial drain on the organization. Consequently, SIPI substantially restructured itself at its annual meeting in 1971. Non-scientists were admitted on a dues-paying basis. The board of directors now included non-scientists with ties to sources of money. A new director was hired and a new staff position—a field organizer to mobilize scientists—was created. As to the relationship between the national and the local organizations, a loose tie was preferred. The journal *Environment* was sold in 1972. SIPI published a series of "workbooks" written by expert scientists on environmental issues. The workbooks became very popular. SIPI also mailed information to journalists upon request and provided seminars for Congressional representatives. In the late 1970s, Commoner increasingly entered the political arena and ran for U. S. President in 1980, whereas Alan McGowan, who took over from Commoner in the SIPI leadership, re-emphasized that the organization should simply provide "neutral facts" (ibid.: 226). Under McGowan, a wide variety of educational programs, with educational grants from foundations, were carried out. In response to the 1979 Three Mile Island accident and the ensuing journalistic demands for information about nuclear power, SIPI set up its Media Resource Service (MRS), a referral service that puts journalists in touch with scientists, functioning as a "giant Rolodex" (ibid.) of 20,000 scientists. This was hugely successful and became the center of SIPI activity. SIPI developed into a national organization, headquartered in New York, with no local affiliates. In the 1980s, it had about 15 staff members and an annual budget of about $2.5 million. It is now defunct, but the Media Resource Service (MRS) prospers. Since 1994, it has been run as a free-of-charge professional service by Sigma Xi, the Scientific Research Society (a supra-disciplinary scientific association), with funding from major foundations, corporations, and media outlets.

References

Moore, K. 1993. Doing Good While Doing Science: The Origins and Consequences of Public Interest Science Organizations in America, 1945–1990. Dissertation, University of Arizona.

Nichols, D. 1974. The associational interest groups of American science. In *Scientists and Public Affairs*, ed. A. Teich. MIT Press.

Society for Advancement of Chicanos and Native Americans in Science

This group was founded 1973 with the purpose of "encouraging Chicano, Latino and Native American students to pursue graduate education in order to obtain the advanced degrees necessary for research careers and science teaching professions."

Current status and activities While SACNAS is a national organization, its membership is geographically concentrated in the southwestern US. Most of the members are science professors. SACNAS holds annual national conferences that are intended to inspire and encourage undergraduates to enter graduate school. The Society also conducts workshops for pre-college science teachers, titled "Teaching Culturally Relevant Science."

Web address http://caldera.calstatela.edu/sacnas/

Mailing address University of California, 1156 High Street, Santa Cruz, CA 95064

Phone number 408 459 4272

Fax number 408 459 3156

E-mail address sacnas@cats.ucsc.edu

Society for Community Research and Action

Purpose "The Society is devoted to advancing theory, research and social action to promote positive well-being, increase empowerment, and prevent the development of problems of communities, groups and individuals." The goals of the SCRA are to promote the use of social and behavioral sciences for the well-being of people and their communities; to promote theory development and research that increase our understanding of human behavior in its social context; and to encourage the exchange of knowledge and skills in community research and action.

History Founded in 1967, the society acquired its current name in 1990.

Current status and activities The Society for Community Research and Action is Division 27 (Community Psychology) of the American Psychological Association. It publishes *The Community Psychologist* (published up to five times a year) and the bimonthly *American Journal of Community Psychology*, selects SCRA fellows, and presents awards, including the Distinguished Contribution to Theory and Research in Community Psychology award, the Distinguished Contribution to Practice in Community Psychology award, the award for Outstanding Contributions to the Mentoring of Ethnic Minority Community Psychologists, and two dissertation awards. The SCRA governance structure includes a variety of committees (e.g., committees focused on women, cultural and racial affairs, social policy, and international concerns) and interest groups (areas of focus include people with disabilities; self-help and mutual support; students of color; rural communities; stress and coping; and lesbian, gay, bisexual and transgender concerns). Current activities include a multi-year interdivisional task force on promoting strengths-based policies and research focused on children, families and communities; developing collaborative activities with other organizations; and a minority pipeline initiative. SCRA also hosts a three-day biennial conference.

Web address http://www.apa.org/divisons/div27

Mailing address 1800 Canyon Park Circle, Building 4, Suite 403, Edmond, OK 73013

E-mail address scra@telepath.com

Society for Freedom in Science

This group was founded in 1940 in the United Kingdom to counteract the influence of a school of thought, led by scholars such as J. D. Bernal and B. Hessen, that was inspired by Soviet science policy. The Bernal group propagated tight governmental control of scientific research, which was to serve immediately related practical ends. At the organizational level, the Society for Freedom in Science opposed the Association of scientific Workers (AscW), which had taken up the Bernal line. The Society's founders were J. R. Baker, Michael Polanyi, and A. G. Tansley.

Purpose To protect the independence of scientific pursuits from external control.

History Because the Society's point of view went counter to the prevailing opinion, the members found it very difficult to get their views published. This led, at the end of the war, to the inauguration of a series of Occasional Pamphlets. Soon the tide began to turn in the Society's favor. As its goals became more widely recognized in the scientific community, its membership dropped, because fewer scientists felt it necessary to become engaged. The Society for Freedom in Science also had members in other countries, notably in the US. The physicist P. W. Bridgman was one of them and also served on the Society's governing Committee.

Reference Society for Freedom in Science. 1953. The Society for Freedom in Science: Its Origins, Objects and Constitution.

Society for Social Responsibility in Science

Viktor Paschkis, who had immigrated to the US from Austria, was the driving force in founding this organization. Paschkis, an engineer, had been offered a position on the Manhattan Project, but turned it down because of his pacifist Quaker worldview. At Columbia University he refused to build an analog computer until the school agreed not to use it for military research. SSRS was formed at a 1948 meeting of scientists at Haverford College. Among the original members were Albert Einstein and Linus Pauling.

Purpose The members of SSRS pledged that they would refuse to participate in research with direct military implications.

History The organization had fewer than 1,000 members (Nichols 1974: 126). In the 1960s, it took a strong stance against the Vietnam War and, in 1969, sent a research team to Vietnam to study the effects of defoliants (before the AAAS-Meselson effort). SSRS became defunct after Paschkis left the organization. However, a German branch, the Gesellschaft für Verantwortung in der Wissenschaft (GVW), founded in 1966, still exists.

References

Inside CPSR. 1991. CPSR Newsletter 9, no. 3 [http://www.cpsr.org/publications/newsletters/issues/1991/Summer1991/inside.html]

Gottfried, K. 1999. Physicists in politics. *Physics Today*, March: 42–48.

Nichols, D. 1974. The associational interest groups of American science. In *Scientists and Public Affairs*, ed. A. Teich. MIT Press.

Smith, A. 1965. *A Peril and a Hope*. University of Chicago Press.

GVW Web address http://mailer.uni-marburg.de/~gvw

GVW mailing address Goldsternstrasse 8, 04329 Leipzig, Germany

GVW phone +49 341 2518841

GVW e-mail gvw@mailer.uni-marburg.de

Society for the Psychological Study of Social Issues

This group was founded in 1936, largely in response to the lack of employment and public policy opportunities for psychologists (especially younger psychologists) during the Depression. Consisting of psychologists, related social and behavioral scientists, students, and others who have an interest in "research on the psychological aspects of important social issues," SPSSI is an international organization that "seeks to bring theory and practice into focus on human problems of the group, the community, and nations, as well as the increasingly important problems that have no national boundaries."

Current status and activities With more than 3,500 members, SPSSI is an independent society as well as Division 9 of the American Psychological Association (APA) and an organizational affiliate of the American Psychological Society (APS). Joining SPSSI does not require membership in either the APA or APS. In addition, the SPSSI has NGO status and serves as a consultant to the Economic and Social Council of the United Nations. SPSSI activities include the publication of the *Journal of Social Issues* and the SPSSI Newsletter, as well as programming at APA's Annual Convention and the Society's own conventions. SPSSI sponsors several award and grant programs, theme conferences, and research workshops. Besides its scholarly endeavors, SPSSI is also active in public policy and has created positions in this area. The James Marshall Public Policy Scholar and the Scientist in the Public Interest work to implement SPSSI public policy initiatives at the federal level. SPSSI graduate student members have full voting privileges and a Student Committee, whose Chair is elected annually and attends SPSSI Council meetings.

Web site http://www.spssi.org

Mailing address P.O. Box 1248, Ann Arbor, MI 48106-1248

Phone number 734 662-9130

Fax number 734 662-5607

E-mail address spssi@spssi.org

Society for the Study of Peace, Conflict and Violence

Purpose "To increase and apply psychological knowledge in the pursuit of peace. Peace here is defined broadly to include both the absence of war and the creation of positive social conditions which minimize destructive conflicts and promote human well-being." SSPCV's goals are defined as follows: "(1) to encourage psychological research, education, and training on issues concerning peace, nonviolent conflict resolution, reconciliation, and the causes, consequences and prevention of war and other forms of destructive conflict; (2) to provide an organization that fosters communication among researchers, teachers, and practitioners who are working on peace issues; (3) to apply the knowledge and the methods of psychology in the advancement of peace, non-violent conflict resolution, reconciliation, and the prevention of war and other forms of destructive conflict."

Current status and activities The Society for the Study of Peace, Conflict and Violence, founded in 1990, is Division 48 of the American Psychological Association (APA). It is a growing group of psychologists, professional affiliates from a variety of disciplines, and students who all share the interest of promoting peace from the small scale of the family to the enormous one of the world at large. The Division publishes the quarterly *Peace and Conflict* and a newsletter. It welcomes all people who wish to contribute to peace, and invites them to contribute "diverse international perspectives and ethnic and gender orientations to peace." Currently it has established ten Working Groups.

Web address http://moon.pepperdine.edu/~mstimac/Peace-Psychology.html

Mailing address 750 First Street, NE, Washington, DC 20002-4242

Phone number 202 336 6013

Fax number 202 218 3599

E-mail address kcooke@apa.org

Society of Hispanic Professional Engineers

In 1974 a group of engineers employed by the City of Los Angeles founded this society as a national organization of professional engineers to serve as role models in the Hispanic community.

Purpose "SHPE promotes the development of Hispanics in engineering, science and other technical professions to achieve educational excellence, economic opportunity and social equity. . . . We will fulfill our mission by increasing educational opportunities, promoting professional and personal growth, carry out our social responsibility to be involved in education, business and government issues and enhancing pride within our organization and reinforcing our reputation as a vital Hispanic organization."

Current status and activities SHPE, Inc., is a national not-for-profit organization, with more than 5,300 student members in 170 student chapters, and more than 1,500 professional members in 42 professional chapters. The SHPE Foundation provides scholarships for engineering and science majors. For its professional members, SHPE organizes seminars on professional growth and career development, and promotes awareness of job opportunities. A major event is the annual National Technical and Career Conference (NTCC). The Society publishes the quarterly *SHPE Magazine*.

Web address http://www.shpe.org

Mailing address 5400 E. Olympic Blvd., Suite 210, Los Angeles, CA 90022
Phone number 323 725 3970
Fax number 323 725 0316
E-mail address shpenational@shpe.org

Standing Committee on Scientific Freedom and Responsibility (American Association for the Advancement of Science)

In 1976, the council and the board of directors of the American Association for the Advancement of Science jointly established the CSFR. In drafting the Committee's Charter, the Committee on Council Affairs recognized that the "increasing interaction among science, technology, and the public interest is bound to continue to raise new issues and problems of professional rights, ethics, and accountability, and the decision rules which worked in the past may no longer suffice."

Purpose Its current charter (most recently revised in 1997) charges CSFR with the tasks, among other things, to encourage and assist AAAS as well as other organizations in developing rules of scientific conduct, to develop policies on scientific responsibility and freedom, and to monitor possible restrictions of scientific freedom worldwide.

History Within AAAS there has been a long-standing concern for the social implications of science, reaching back to the 1930s. A major milestone was reached with the Arden House Statement of 1951, in which the Executive Committee of AAAS urged a stronger emphasis on "the relations of science to government, and indeed the relations of science to our society as a whole." A few years later, the string of important AAAS committees that have addressed various aspects of the science-society interface began. Worried about nuclear radiation hazards and chemical pollutants, such as DDT, biochemist Ward Pigman asked the AAAS to set up a committee on the societal impacts of science. In 1957, the interim *AAAS Committee on the Social Aspects of Science* was founded, with Pigman as its first chair. Among its proposals was the establishment of a Committee on Science in the Promotion of Human Welfare as its permanent successor. Such a committee was indeed instituted in 1958, and Barry Commoner became its chair. It produced a number of reports, among them the 1965 report on "The Integrity of Science." When Margaret Mead took over as chair in 1966, work on the race issue moved to the foreground. A Council Study Committee on Ethics was appointed in 1964. Starting out in 1966, an informal Science and Public Policy Study Group became instrumental in helping academic programs on science and technology policy establish themselves. In 1970, AAAS, in collaboration with NAS, undertook a study (headed by Matthew Meselson) of the effects of defoliants in Vietnam. Simultaneously with the establishment of the CSFR, there was a push within AAAS to revise its constitution. In 1977, the AAAS Constitution was amended so that the AAAS objectives included "to foster scientific freedom and responsibility, to improve the effectiveness of science in the promotion of human welfare."

Current status and activities CSFR has strong ongoing activities related to human rights and professional ethics. It publishes a quarterly newsletter, *Professional Ethics*

Report, and a *Report on Science and Human Rights*. The Scientific Freedom, Responsibility and Law Program and the Science and Human Rights Program of AAAS have also issued a series of topical papers, reports, and books (see second and third Web sites listed below).

Sources

AAAS Committee on Science in the Promotion of Human Welfare. Science and Human Welfare. *Science* 132 (July 8, 1960): 68–73.

Scientific Freedom and Responsibility. 1975. A Report of the AAAS Committee on Scientific Freedom and Responsibility ("Edsall Report"). AAAS.

Frankel, M. S., and Teich, A. H., eds. 1994. *The Genetic Frontier*. AAAS.

Wolfle, D. L. 1989. *Renewing a Scientific Society*. AAAS.

Web address

http://www.aaas.org/

http://shr.aaas.org/program/index.htm

http://www.aaas.org/spp/dspp/sfrl/sfrl.htm

Mailing address 1200 New York Ave. NW, Washington, DC 20005

Phone number 202 326 6793

Fax number 202 289 4950

E-mail address mfrankel@aaas.org

Stockholm International Peace Research Institute

The idea of a peace research institute was first put forward by Swedish Prime Minister Tage Erlander in 1964. A Swedish Royal Commission chaired by Alva Myrdal (1982 Nobel Peace Prize Laureate) recommended its establishment in 1966. In the same year, the Swedish Parliament created SIPRI as an independent foundation.

Purpose SIPRI's mission is to "conduct scientific research on questions of conflict and co-operation of importance for international peace and security, with the aim of contributing to an understanding of the conditions for peaceful solutions of international conflicts and for a stable peace." The Institute's work is intended to be of "an applied research character directed towards practical-political questions [which] should be carried on in a constant interchange with research of a more theoretical kind."

Current status and activities Major current research areas include conflicts and conflict prevention, management and resolution; arms control issues; arms transfers; arms production; military expenditure; military technology; chemical and biological weapons; European security; and export controls. SIPRI maintains security-related databases, organizes conferences, and issues a wide range of publications, among them the *SIPRI Yearbook: Armaments, Disarmament and International Security*, to disseminate its research findings.

Web address http://www.sipri.se/

Mailing address Signalistgatan 9, S-16970 Solna, Sweden
Phone number +46 8 6559700
Fax number +46 8 6559733
E-mail address sipri@sipri.se

Union of Concerned Scientists

By the end of 1968, a deep apprehension had grown among MIT faculty and students about the Vietnam War and about what they considered abuses of science and technology by the government. To "protest the misuse of science and technology, especially as it related to defense research at universities" (UCS 1993: 1), a group of MIT faculty and students organized a one-day research stoppage at MIT for March 4, 1969. The steering committee for this event picked "Union of Concerned Scientists" as its name. Students and faculty soon split, because the students' goals and actions were far more radical than the faculty's. The faculty group continued as UCS, which soon attracted a wider membership and became independent of MIT. MIT physics professor Herman Feshbach was the first chair of UCS.

Purpose UCS aims at the impartial use of scientific knowledge and expertise in the service of solving social problems. It has been dedicated to "advancing responsible public policies in areas where technology plays a critical role" (UCS 1993: 1). "We carry out this mission through technical studies and public education, and by working to influence government policy at the local, state, federal, and international levels." (ibid.)

History UCS started as a small, informal organization, funded by small donations from many individuals. The members acted on a case-by-case basis, and there was intermittent contact with other organizations. Although the original catalyst of UCS was the Vietnam War and other military issues, energy and environmental topics soon came to the fore. In 1970 the UCS Committee on Environmental Pollution (UCS-CEP) was formed, which produced reports on a variety of environmental problems. UCS experts presented testimony at the Boston Area Air Pollution Hearings in 1970, for which they received widespread support in the community (Moore 1993: 207). A year later, UCS published its first major safety report on nuclear power, an area that was to become a major focus of UCS's work. The visibility and credibility UCS achieved during the 1970s through its activities on this issue allowed the organization to acquire a full-time staff, with offices in Cambridge, and later also in Washington and Berkeley (Gottfried 1999). In the 1980s, arms control became a major theme (Casper and Krauss 1981). UCS, for instance, lobbied against the MX missile and the Strategic Defense Initiative, and supported the START treaties. By the end of the 1980s, UCS increasingly turned to the threat of global warming and other environmental issues, and greatly expanded its work on renewable energy. The Sound Science Initiative, launched in 1995, includes an electronic alert network that helps participating scientists raise public awareness and influence policies on global warming, biodiversity loss, and the impacts of population growth. A UCS-sponsored statement "World Scientists Call For Action at the Kyoto Climate Summit" was endorsed by 109 Nobel Laureates and more than 1,500 national academy-level scientists from around the world. By the middle of the 1980s, UCS acted both as a "watchdog" group and as a trusted political insider who was called upon to participate in the government decision-making process (Moore 1993: 214). Until his death in 1999, physics Nobel laureate Henry Kendall, of MIT, was the major driving force behind UCS; he was chairman for twenty-five years. Kurt Gottfried, physicist and co-founder of UCS, succeeded him in the chair.

Current status and activities UCS is now a broad "multi-issue organization (H. Ris in UCS 1993: 12)," with programs in five areas: global resources, sustainable agriculture, energy, transportation, and arms control. It has a staff of about 50 in three offices (Cambridge, Berkeley, and Washington), and a contributing sponsorship of about 70,000 (consisting of both scientists and non-scientists). It is financially sound, with total annual operating revenue of about $5.0 million for 1997. Its fundraising and lobbying efforts are of professional quality. UCS issues a quarterly magazine, titled *Nucleus*.

References

Allen, J., ed. 1970. *March 4*. MIT Press.

Balogh, B. 1991. *Chain Reaction*. Cambridge University Press.

Casper, B., and Krauss, L. 1981. Fortune favors the prepared mind—a movement against nuclear war. *Science, Technology, and Human Values* 7, fall: 20–26.

Gottfried, K. Physicists in politics. *Physics Today*, March 1999: 42–48.

Kendall, H. 2000. *A Distant Light*. AIP and Springer Verlag.

Moore, K. 1993. Doing Good While Doing Science: The Origins and Consequences of Public Interest Science Organizations in America, 1945–1990. Dissertation, University of Arizona.

UCS. 1993 Annual Report.

UCS. 1997 Annual Report.

Web address http://www.ucsusa.org/

Mailing address Two Brattle Square, Cambridge, MA 02238-9105

Phone number 617 547 5552

Fax number 617 864 9405

E-mail address ucs@ucsusa.org

Women in Engineering Programs and Advocates Network

Founded in 1990 to be a "catalyst for change that enhances the success of women in the engineering professions," this non-profit 501(c)(3) educational organization is headed by a 23-person board of directors from academe and industry and has more than 500 members. It has raised more than $3.5 million in support of its activities, which include technical assistance, training, research, and publications (among them *WEPANews*). WEPAN organizes national conferences and presents awards. Among WEPAN's major initiatives was a "Climate" Survey about gender differences among undergraduate engineering students. A curriculum for training mentors and mentees in science and engineering was developed. Furthermore, since

1997, MentorNet has paired students with professionals in industry for year-long mentoring relationships via e-mail.

Web address http://www.wepan.org

Mailing address 1284 CIVL Building, Room G293, West Lafayette, IN 47907-1284

Phone number 765 494 5387

Fax number 765 494 9152

E-mail address wiep@ecn.purdue.edu

Working Group on Ethical, Social and Legal Implications (Human Genome Project)

The leaders of the Human Genome Project recognized the need to study its societal implications. "One of [James Watson's] first acts as director of the Human Genome Project was to announce that 3 percent (now 5 percent) of its funds would be set aside for studies of its ethical and legal consequences." (*New York Times*, April 7, 1998) At its January 1989 meeting, the Program Advisory Committee on the Human Genome established a working group on ethics to develop a plan for the ELSI component of the overall project. In response to the working group's report, the National Human Genome Research Institute established the ELSI Branch (later renamed the ELSI Research Program) in 1990. The Office of Energy Research of the Department of Energy, which participates in the US Human Genome Project, also reserved a portion of its funding for ELSI research.

Purpose To study the societal implications of the Human Genome Project.

History The four original priorities of ELSI were privacy and fair use of genetic information; clinical integration of genetic information into medicine; future research issues; and education. In the fall of 1998, a year-long evaluation and review process of the ELSI program was concluded. The review resulted in five new goals (listed below).

Current status and activities ELSI is administered by the National Human Genome Research Institute (NHGRI), which belongs to the National Institutes of Health (NIH). NHGRI reserves five percent of its annual research budget for ELSI projects. By the end of 1998, the ELSI programs had funded almost $40 million worth of research. As stated on its Web site, its five priority areas for 1998–2003 are as follows: (1) Examine the issues surrounding the completion of the human DNA sequence and the study of human genetic variation. (2) Examine issues raised by the integration of genetic technologies and information into health care and public health activities. (3) Examine issues raised by the integration of knowledge about genomics and gene-environment interactions into nonclinical settings. (4) Explore ways in which new genetic knowledge may interact with a variety of philosophical, theological, and ethical perspectives. (5) Explore how socioeconomic factors and concepts of race and ethnicity influence the use, understanding, and interpretation of genetic information, the utilization of genetic services, and the development of policy.

Reference Rapp, R., Heath, D., and Taussig, K. S. Tracking the Human Genome Project. *Items* 52 (1998), no. 4: 88–91.

Web address http://www.nhgri.nih.gov/elsi/

Mailing address National Institutes of Health, Building 31, Room B2B07, 31 Center Drive, MSC 2033, Bethesda, MD 20892-2033

Phone number 301 402 4997

Fax number 301 402 1950

E-mail address elsi@nhgri.nih.gov

World Federation of Scientific Workers

The parent organizations of this group were the British Association of Scientific Workers and its sister organizations in other countries, such as the American Association of Scientific Workers and the French Association des travailleurs scientifiques. At a conference in July 1946 in England, a number of such national scientific associations formed the World Federation of Scientific Workers. Among the founding members were J. D. Bernal and Frederic Joliot-Curie.

Purpose To promote the utilization of science for peace and for the welfare of humankind, to further the international exchange of scientific personnel and basic research information, to improve the status of scientific workers, and (according to the preamble to the federation's constitution) "to encourage scientific workers to take an active part in public affairs, and to make them more conscious of, and more responsive to, the progressive forces at work within society."

History In comparison with the Federation of American Scientists, WFScW included a stronger trade union element—improving the working conditions and status of scientific workers—in its goals, and it was also more left-leaning. For these reasons, FAS declined to affiliate with the World Federation of Scientific Workers.

Current status and activities The World Federation of Scientific Workers's current main themes include science and society, science policy, disarmament, science and ethics, science and the law, science and human rights, women scientists, young scientists, as well as scientific co-operation with poor and developing countries, especially in conjunction with UNESCO.

Reference Smith, A. 1965. *A Peril and a Hope*. University of Chicago Press.

Web address http://assoc.wanadoo.fr/fmts.wfsw

Mailing address Case 404, 93514 Montreuil, Cedex, France

Phone number +33 1 48188175

Fax number +33 1 48188003

E-mail address fmts@wanadoo.fr

Notes

Chapter 1

1. See also Shadish 1989.

2. The current jargon of science policy routinely pairs science and technology, yet one should not forget that this pairing is relatively new (de Solla Price 1984; see also, more generally, de Solla Price 1961). Science and technology developed quite independently during the early phases of the industrial revolution—most of the key technological breakthroughs of that era came out of a practical "tinkering" tradition rather than out of academic science. But from the nineteenth century on, the interpenetration between science and technology and that between these two areas and the economic domain have increased, and this has enormously accelerated technological progress and economic prosperity (Branscomb and Keller 1998; Hart 1998; Gibbons 1997; Solow 1957).

3. Brooks is commonly misquoted as having distinguished "policy for science" and "science *for* policy."

4. See, e.g., Branscomb and Auerswald 2001; Griliches 1998; Hart 1998; Mowery and Rosenberg 1998; Nelson 1996; Nelson and Phelps 1965; Rosenberg 1972. For a large collection of data on this topic, see NSB 2000.

5. The second edition of the *Oxford English Dictionary* lists this author's work as the earliest reference. Sainte-Beuve's *Pensées d'août, Poésies* (1837) contain a poem titled "Á M. Villemain" with the lines "(Il a maintient encore;) et Vigny, plus secret / Comme en sa tour d'ivoire, avant midi, rentrait" ["And Vigny, more secretive, retreated before noon, as if to his ivory tower"]. In the 1850s, John Henry Newman's lectures, published under the collective title "The Idea of A University," closely linked the metaphor to the university. *Ivory Tower* was also the title of an unfinished novel by Henry James.

6. A recent electronic keyword search for "ivory tower" in the Harvard library system retrieved 83 items. Interestingly, a good number of these publications are variations on a "beyond the ivory tower" theme. Bok 1982 and Zuckerman 1971 include this very phrase in their titles.

7. For detailed discussions of this intellectual movement, see Holton 1993 and Koertge 1998. See also Segerstråle 2000.

8. For a critical discussion of this school of thought, see Bricmont and Sokal 2001.

9. One who opposed this view of science was Eric Ashby (1971: 32), who exclaimed in his 1971 Bernal Lecture: "We should not leave unchallenged those who wish to

impose politically directed goals on basic (or, as Americans call it, 'discipline-oriented') research." According to Ashby, reducing science to a socially useful enterprise may paradoxically even intensify anti-science sentiments in the population.

10. Source: Pugwash Web site. See also Rotblat 1999. One half of that year's Nobel Peace Prize went to Rotblat in person, the other half to the Pugwash organization.

11. The relationship between science and society has varied across countries. On the Central European tradition, for example, see Ringer 1990.

12. To prevent the concentration of power that might lead to new kinds of non-accountable establishments, the framers of the constitution took care to decentralize power both within government (through a system of checks and balances) and between government and civil society (through the protection of civil liberties).

13. On a much smaller scale, the federal government was involved in scientific research even before World War II. A large part of this research was in agriculture, in connection with the Land Grant scheme. For a more detailed historical survey of the government's science activities, see Sonnert and Brooks 2001.

14. Source: *American Institute of Physics Bulletin of Science Policy News*, January 10, 2000. This quote is attributed to Representative Rush Holt (D-New Jersey), who was a physicist before entering politics.

15. "Bureaucrat-scientist," which we considered using here, has the advantage of paralleling "citizen-scientist." However, we decided against using that term because it might be misunderstood as having a derogative intent.

16. The distinction is similar, though not identical, to Merton's (1968) distinction of bureaucratic and unattached intellectuals. Merton restricts his definition of intellectuals to the fields of social science; and his group of unattached intellectuals includes all those not involved in policy making through working in a bureaucracy. The citizen-scientists, in contrast, are individuals from all scientific fields who seek to influence policy by organizing.

17. Science advising has attracted the attention of scholars of different schools ever since it was instituted. (See, e.g., Abelson 1965; Brooks 1964; Gilpin and Wright 1964; Graham 1991; Hart 1998; Jasanoff 1990; Mullins 1981; Perl 1971; Smith 1992.) The connection between the Cold War and the construction of the first ivory bridge was studied by Leslie (1993), by Lowen (1997), and by Wang (1999). Kevles (1987) covered parts of the topic in his disciplinary study of the physics community, which was very much in the forefront of the evolving system of science policy, especially in the early years. For a wider historical perspective, see Weart 1979.

18. We will use *Jeffersonian research* and *Jeffersonian Science* as synonyms.

19. Two major contributions to this debate from the political sector were Clinton and Gore 1994 and House Committee on Science 1998.

20. In more recent years, the science policy system has increasingly reached even beyond the national framework through international science collaborations and organizations, such as the Intergovernmental Panel on Climate Change, the International Institute for Applied Systems Analysis, and the International Science and Technology Center (all described in appendix D). The political impact of international scientific cooperation on nations in conflict was the topic of a conference

held in January 1998 at the New York Academy of Sciences and of a conference held in February 2000 at Futuribles International in Paris.

21. Merton's postulate of a truth-based ethos of science became a major target for critiques by relativist or constructivist sociologists of science (e.g., Collins 1981; Latour and Woolgar 1979). The fact that scientists sometimes follow norms that contradict Merton's ethos of science (Mitroff 1974) was of course no surprise to Merton himself. Such deviance does not invalidate this ethos as the officially and structurally defining rationale of the scientific subsystem. Moreover, most scientific practitioners still appear to hold onto the non-relativist theory of knowledge spelled out in the ethos, sociological critiques notwithstanding.

22. Price's (1965) other two estates were the professional and the administrative. These two intermediary estates between science and politics include engineers, medical professionals and applied scientists (professional estate) and government administrators (administrative estate), and together roughly correspond to the "scientist-administrators." None of the estates covered the "citizen-scientists."

Chapter 2

1. For instance, in 1999, according to the Science and Engineering Indicators (NSB 2000: A-565), 47% of American adults believed that the benefits of scientific research strongly outweighed harmful results, and an additional 27% thought that the benefits slightly outweighed the harmful results. In a study of Michigan residents' attitudes toward medical research, which was conducted in early 1998 by Charlton Research Company for Research!America, 82% of the respondents believed that, even if it brought no immediate benefits, basic science research should be supported by the federal government. The same proportion of respondents expressing this opinion was found by the National Science Foundation in 1999 (ibid.: 8–15, A-560). See also Miller 1999, 2000.

2. For example, under the leadership of Representative Vernon Ehlers, the House Committee on Science produced a report to Congress, entitled "Unlocking Our Future: Toward a New National Science Policy" in September 1998. For a scholarly discussion of the issue, see Guston 2000a,b.

3. For a well-reasoned recent argument in support of basic research, see Committee for Economic Development 1998.

4. For a detailed discussion of the various rationales on which US science and technology policy has been based at various times, see Crow and Bozeman 1998.

5. This argument is spelled out in detail in Holton, Chang, and Jurkowitz 1996. Ehrenreich 1997 is also instructive.

6. Branscomb (1998) argues that the meaning of "applied research" has become too fuzzy; he suggests "problem-solving research" as a replacement.

7. Our term *Baconian science* thus derives from Bacon's respect for the applications of science, whereas other authors sometimes use the word Baconian to refer to a different aspect of Bacon's complex ideal of science: theory-free observation.

8. For details of Bush's career and views, see Zachary 1997.

9. Among the chief antagonists of Bush's efforts to protect basic research was Senator Harley Kilgore (D-West Virginia). Kilgore insisted that science should be more tightly harnessed to addressing society's needs. His views were an exposition of what Lewis Branscomb, in a personal communication, once called the "use it or lose it" philosophy of science management.

10. See Holton 1986, 1988, 1998; Holton and Sonnert 1999.

11. Elsewhere (e.g., Nichols 1986), the word *Jeffersonian* has sometimes been used to describe a decentralized approach.

12. Branscomb further emphasizes correctly that the motivation of the research sponsor is not necessarily identical with the motivation of the researcher. Researchers typically seem less interested than their sponsors in the practical benefits of their work. They tend to prize the "basicness" of the research process.

13. It could be argued that, in the two-by-two table implied by Branscomb's scheme, one cell—an "applied" research process with no motivation to gain any concrete benefits—must in practice remain fairly empty.

14. In recounting the history of this endeavor, one must avoid overemphasizing, in hindsight, the grand rationale behind events that were also strongly shaped by pragmatic and political strategies, short-term objectives, and sheer happenstance. Some of the actors stress the importance of day-to-day concerns. In a personal communication, Philip Smith, one of Frank Press's deputies at the time, recalled that a major objective of the OSTP staff was to "continue to sell a somewhat reluctant president on the importance of funding basic science, because of his hostility to universities." The second immediate objective was, according to Smith, to shore up the crumbling base for fundamental research across the different departments and agencies of federal government.

15. For an instructive overview of the major trends in the US research system up to the mid 1980s, see Brooks 1985.

16. For a brief history of presidential science advising, see appendix A.

17. See also Press 1978b, 1995.

18. The sources of information gathered by these methods are cited as personal communications.

19. Within the National Academy of Sciences, the task of producing this report, as well as the other NAS reports mentioned, was assigned to the Committee on Science and Public Policy, which was chaired during the first phase of this work, until 1965, by George Kistiakowsky, and then, until 1971, by Harvey Brooks.

20. In 1969, a third report, titled Technology: Processes of Assessment and Choice, was issued. This report, which was beyond the scope of our present topic, became the blueprint for the Office of Technology Assessment.

21. See also Strickland 1989: 69–70.

22. See also Greenberg 1966.

23. See, e.g., *Science* 154 (1966): 1123; *Science* 155 (1967): 150; *Science* 157 (1967): 1512. In 1968, an NSF-sponsored report titled Technology in Retrospect and Critical Events in Science, came to quite opposite conclusions (Illinois Institute

of Technology Research Institute 1968). Examining the history of five breakthrough innovations, it found that, in each case, about 70% of the important papers that contributed to one of these innovations could be classified as reports of basic research. Ten years later, during the Carter administration, a medical study (Comroe and Dripps 1978) similarly supported the view that basic research plays a crucial role in propelling applied technology. This study examined the top ten advances of cardiovascular and pulmonary medicine and surgery between 1945 and 1975 and found that 42% of the research articles that contained key contributions to these advances reported research whose goal was unrelated to those later clinical breakthroughs. For one of the many studies on this subject, see Holton, Chang, and Jurkowitz 1996.

24. Some of Carter's animosity toward colleges and universities might be traced to Admiral Hyman Rickover, who was in charge of the nuclear submarine program in which Carter served. According to Harvey Brooks (personal communication), Rickover "was extremely antagonistic to universities and to academic scientists in general, and was particularly allergic to advisory committees of all sorts."

25. In making this point, Press referred to the central argument that the scientific community used for justifying the expenditure of taxpayers' money on basic research—that of the "unpredictable spinoff": Advances in basic research would lead to substantial practical benefits in unpredictable ways.

26. An amendment to the Defense Procurement Authorization Act of 1970 by Senator Mike Mansfield (D-Montana) prohibited the Department of Defense from funding any research not directly related to military function or operation, and directed the NSF to conduct the more basic research programs that up to that point were run by the military (Burger 1980: 107; Fuqua 1993: 413).

27. Reducing the paperwork burden for applicants and grantees in the federal research administration became an important thrust and achievement of the Carter years. Several specific measures, such as limitations on the page length of applications and annual reports, that were inaugurated under Carter, are now embedded in agency practices (Gilbert Omenn, personal communication).

28. According to a memorandum to Press from Dick Meserve (December 29, 1978), the Department of Defense also enthusiastically implemented the recommendations of the Galt Committee Report of May 19, 1978, which aimed at strengthening basic research in that department.

29. The OSTP news release containing the entire list of Jeffersonian-type basic research questions that various government units considered to be in the national interest is presented in appendix C below.

30. A related question emerged from the Department of Defense: "Can we discover anti-viral agents to combat viral diseases? The development of such drugs would have as large an effect on mankind as did the discovery of antibiotics." This demonstrates a certain convergence, at the level of basic research questions, between government agencies whose immediate missions are very different. Similarly, high-temperature superconductivity was brought up, as has been mentioned, by both the Department of Defense and the Department of Energy.

31. A plausible explanation for the Department of Health, Education and Welfare's silence is that this department tended to defer to NIH on nearly all research inquiries.

32. For details of the development of NIH, see Strickland 1989.

33. It is not clear why the Department of Transportation's questions were omitted.

34. The complete list is presented in appendix C.

35. RANN's precursor was a program called Interdisciplinary Research Relevant to Problems of Our Society. After 2 years it was expanded into RANN, one of the most controversial NSF programs ever.

36. On the science policy of the Reagan administration, see Keyworth 1993.

37. In 1982 the National Academy of Sciences' Committee on Science and Public Policy (COSPUP) merged with parallel activities at the National Academy of Engineering to form COSEPUP. The Institute of Medicine became the third sponsor of COSEPUP.

38. In the academic year 2000–2001, together with colleagues from three Harvard faculties and a group of students, we explored this issue under a Harvard University Provost's Grant. From our discussions, it indeed appeared that the perceived irrelevance of basic science to real-life problems might be one of the hindrances to the choice of a science career, especially for women and members of underrepresented minorities.

39. Alternatively, one might think of using also a non-governmental consultative body, along the lines of the "National Forum on Science and Technology Goals," that was among the recommendations of the 1992 Carnegie Commission report Enabling the Future: Linking Science and Technology to Societal Goals, or along the lines of focus groups, such as those assembled by the organization Public Agenda or by the American Association for the Advancement of Science.

Chapter 3

1. Whereas citizen-scientists typically oppose governmental uses of science and technology, they occasionally criticize the government's inaction or neglect in respect to developing scientific possibilities. It might be interesting to follow the burgeoning debate about human stem cell research under this aspect.

2. Of course, some perceptive scientists, beginning with Niccolo Tartaglia in the early sixteenth century, had long anticipated the responsibility to look at the possible misuse of scientific advances. Pierre Curie, speaking in 1903 during his high moment in Stockholm, reflected on the possible consequences of the discovery of radioactivity, worrying whether humankind was "mature enough to take advantage" of such laboratory discoveries.

3. Some associations are in the gray area between public-interest and professional organizations. The Dark Sky Association, for instance, combines the professional interest of astronomers in dark skies with societal concern about light pollution. One might here include also the Creation Research Society (which rejects Darwin's

theory of evolution) and the National Center for Science Education (which defends it), insofar as the controversy between Creationist and Darwinian theories has more than merely science-internal significance.

4. The Rochester Section is apparently identical with the "University of Rochester Atomic Group" listed as a participating organization in an earlier meeting (Smith 1965: 229).

5. At Oak Ridge, personnel was segregated by the various plants and laboratories because these were run by different companies (Smith 1965: 98). Consequently, three organizations—one of scientists, one of production scientists, and one of engineers—were formed independently. (The latter two merged in December 1945.) It is not clear whether a cultural gulf between scientists and engineers was an additional cause of the organizational separateness. Alice K. Smith (1965: 111) commented: "In retrospect, a marked difference between the two groups [scientists and engineers] is apparent, and it was diametrically opposite to what one would, offhand, expect. The scientists, many of them absorbed professionally in highly abstruse theory, swung into action as political tacticians; the engineers, or at least those who took any part in the movement at all, concentrated on longer range solutions and became enthusiastic students of world organization."

6. In the mid 1970s, specific research advances in microbiology (e.g., recombinant DNA), which some considered sources of potential hazards and which brought many biologists to activism, added impetus to the growing emphasis on ecology and the environment (Weiner 1979). In a fascinating development, for which the two Asilomar Conferences (1973 and 1975) are memorable milestones, leading researchers in the areas of tumor viruses and recombinant DNA collectively assessed the risks of their newly developed techniques and produced guidelines that provided a blueprint for subsequently issued NIH official guidelines. In this famous case of self-regulation, the scientists themselves took the lead and the federal science apparatus followed.

7. An earlier organization in this area was the Coalition for Responsible Genetic Research, founded in 1976.

8. One gets into murkier territory when one asks about the global effect scientists' public-interest associations have had on the status of science in society. This question will be addressed below.

9. In his classic treatise on rational choice, Mancur Olson (1965) discussed the conditions for collective action. A central problem within the rational-choice framework is that of the "free rider," which comes into play when the benefits of collective action accrue to all, including those who did not incur the expenses of activism. Why, then, should people join social movements? It may be possible to modify the original rational-choice assumptions in a way that would explain movement participation within that general theory (Mitchell 1979). Moreover, members of social movements rarely view their participation instrumentally within a cost-benefit accounting, but they often seem driven by an idealistic belief in the importance of the cause itself. They are typically motivated by a sense of shared fate (Schwartz and Paul 1992). To some extent, participation is not only a means to an end but also its own reward, because it generates satisfying experiences of solidarity and

community. Clarence Lo (1992) argued that especially radical groups who challenge the societal status quo spring from the mobilization of communal identities rather than from the individuals' cost-benefit accounting. In sum, social-psychological approaches may usefully complement the theories of social-movement formation that derive from a rational-choice framework (Gamson 1992).

10. That intermediate plane between the societal level and the individual level has been the focus of resource-mobilization theory, a well-established theory for studying social movements (McAdam, McCarthy, and Zald 1988).

11. On the range of tactics and strategies used by social-movement organizations, see also Barkan 1979; Haines 1984, 1988; McAdam 1983.

12. The term "military-industrial complex" is widely credited to President Eisenhower, who used it in his farewell address. Later it became a favorite of the left and was expanded to "military-industrial-academic complex."

13. As we have already pointed out, even this "radicalism" would be considered rather tame in a wider perspective. Perhaps contrary to wishful views of the workings of a pluralist democracy, some sociological studies found that the use of disruptive and even violent tactics was positively related to a group's political effectiveness (Gamson 1975; see also Tarrow 1994). Moreover, in what is known as a positive "radical flank effect" (Haines 1988), the existence of radical groups may benefit the movement as a whole. Haines's (1984, 1988) work on the civil rights movement showed that the activities of extremists boosted funding for the moderates, who started looking respectable by comparison, and appeared to be a bulwark against the wholesale radicalization of the movement. Yet, on the other hand, a negative backlash is also a possible reaction to the presence of a radical wing in a movement (Barkan 1986).

14. Of course, many of the UCS "moderates" were veterans of the radical leftist politics in the 1930s who by the end of the 1960s had achieved very senior and respectable positions within academic science.

15. See also Perrow 1970.

16. A classic example is that of the March of Dimes (Sills 1957). Founded in the 1930s with the specific mission of fighting poliomyelitis, that association did not dissolve after a vaccine developed by Jonas Salk wiped out this disease; instead it turned to combating birth defects. The Citizens' League Against the Sonic Boom is an example of an exception. Formed with a concrete and limited goal—preventing a large presence of supersonic transport airplanes—it promptly disbanded once that goal was accomplished (Hedal 1980; Shurcliff 1970). In this simple but rare pattern, success terminates the organization.

17. The FAS was not founded by individual scientists; it was a federation of preexisting local scientists' groups. Its tentative original name, Federation of Scientific Organizations, reflected this.

18. Among the radical organizations, Nichols (1974: 126) also mentioned three somewhat obscure groups: HealthPAC, the Physicians Forum, and the Student Health Organization.

19. Source: Natural Resources Defense Council Web site.

20. For more details, see the AAAS-related profiles in appendix D.

21. For a summary of NAS activities, see Halpern 1997.

22. For instance, the Loka Institute, with its emphasis on citizens' involvement and discursive decision making about science, harks back to a central concern of SftP. In view of the sometimes drastic consequences, one might also regard the aggressive litigation strategies pursued by the National Resources Defense Council and similar groups as radical, although these groups operate strictly within the legal status quo and do not express revolutionary aims.

23. The Nobel Peace laureates include Linus Pauling (1962), the International Physicians for the Prevention of Nuclear War (1985), Joseph Rotblat and the Pugwash Conferences on Science and World Affairs (1995), the International Campaign to Ban Landmines (1997), and Médecins Sans Frontières (1999).

24. International treaties must complete a protracted and sometimes hazardous journey before they can take effect. After the treaties have been negotiated and signed by the governments involved, they must be ratified. Ratification cannot be taken for granted; for instance, in 1999 the US Senate refused to ratify a Comprehensive Nuclear Test Ban Treaty.

25. But see, for example, I. I. Rabi's success in this respect (Holton 1999; Rigden 2000).

26. Individuals outside the aggrieved group who work for its cause are known as "conscience constituents" (McCarthy and Zald 1973, 1977).

27. On the career aspects of the gender issues in science, see Sonnert and Holton 1995a,b.

Chapter 4

1. One interesting experimental study investigated how a minimal educational intervention changed the public's views on complex scientific-technological issues (Doble 1995). After the intervention, the members of the public arrived at views very similar to those held by a sample of scientists, who independently also responded to the questions given to the study participants after the intervention.

References

AAAS Committee on Scientific Freedom and Responsibility. 1975. Scientific Freedom and Responsibility. American Association for the Advancement of Science.

Abelson, P. H. 1965. The president's science advisers. *Minerva*, winter: 149–158.

Allen, J., ed. 1970. *March 4: Scientists, Students and Society*. MIT Press.

Ashby, E. 1971. Science and antiscience. *Proceedings of the Royal Society of London* 178: 29–42.

Ausubel, J. H. 1998. Reasons to worry about the human environment. *Cosmos* 8: 1–12.

Baker, W. O. 1993. Notes on science advising in the White House. In *Science and Technology Advice to the President, Congress, and Judiciary*, ed. W. Golden. AAAS Press.

Barber, B. 1953. *Science and the Social Order*. Allen & Unwin.

Barfield, C. E. 1982. *Science Policy from Ford to Reagan: Change and Continuity*. American Enterprise Institute for Public Policy Research.

Barkan, S. E. 1979. Strategic, tactical, and organizational dilemmas of the protest movement against nuclear power. *Social Problems* 27: 19–37.

Barkan, S. E. 1986. Interorganizational conflict in the Southern civil rights movement. *Sociological Inquiry* 56: 190–209.

Barnes, S. H., and Kaase, M. 1979. *Political Action: Mass Participation in Five Western Democracies*. Sage.

Beck, U. 1992. *Risk Society: Towards a New Modernity*. Sage.

Becker, H. S. 1963. *Outsiders: Studies in the Sociology of Deviance*. Free Press of Glencoe.

Beckler, D. Z. 1976. The precarious life of science in the White House. In *Science and Its Public*, ed. G. Holton and W. Blanpied. Reidel.

Belenky, M. F., Clinchy, B. M., Goldberger, N. R., and Tarule, J. M. 1986. *Women's Ways of Knowing: The Development of Self, Voice, and Mind*. Basic Books.

Bell, D. 1973. *The Coming of Post-Industrial Society: A Venture in Social Forecasting*. Basic Books.

Ben-David, J. 1984. *The Scientist's Role in Society: A Comparative Study*. University of Chicago Press.

Berg, P., and Singer, M. 1998. Inspired choices. *Science* 282, October 30: 873–874.

Blanpied, W. A., ed. 1995. *Impacts of the Early Cold War on the Formulation of US Science Policy: Selected Memoranda of William T. Golden, October 1950–April 1951*. AAAS.

Bok, D. 1982. *Beyond the Ivory Tower: Social Responsibilities of the Modern University*. Harvard University Press.

Brand, K.-W. 1982. *Neue soziale Bewegungen: Entstehung, Funktion und Perspektive neuer Protestpotentiale, eine Zwischenbilanz*. Westdeutscher Verlag.

Branscomb, L. M. 1993. Science and technology advice to the US government: Deficiencies and alternatives. *Science and Public Policy* 20, no. 2: 67–78.

Branscomb, L. M. 1995. Public funding of scientific research: Policy criteria for investigator discretion, sponsor's intent, and accountability for outcomes. In *Vannevar Bush II: Science for the 21st Century*. Sigma Xi, The Scientific Research Society.

Branscomb, L. M. 1998. From science policy to research policy. In *Investing in Innovation*, ed. L. Branscomb and J. Keller. MIT Press.

Branscomb, L. M. 1999. The false dichotomy: Scientific creativity and utility. *Issues in Science and Technology* 16, no. 1: 66–72.

Branscomb, L. M., and Auerswald, P. E. 2001. *Taking Technical Risks: How Innovators, Executives, and Investors Manage High-Tech Risks*. MIT Press.

Branscomb, L. M., and Keller, J. H., eds. 1998. *Investing in Innovation: Creating a Research and Innovation Policy That Works*. MIT Press.

Branscomb, L. M., Holton, G., and Sonnert, G. 2001. Science for Society—Cutting-Edge Basic Research in the Service of Public Objectives: A Blueprint for an Intellectually Bold and Socially Beneficial Science Policy. Report on the November 2000 Conference on Basic Research in the Service of Public Objectives. Belfer Center For Science and International Affairs.

Brennan, D., Douglas, W. O., Johnson, L., McGovern, G. S., and Wiesner, J. B. 1969. *ABM, Yes or No?* Center for the Study of Democratic Institutions.

Bricmont, J., and Sokal, A. 2001. Science and sociology of science: Beyond war and peace. In *The One Culture*, ed. J. Labinger and H. Collins. University of Chicago Press.

Brint, S. 1984. "New-class" and cumulative trend explanations of the liberal political attitudes of professionals. *American Journal of Sociology* 90: 30–71.

Bromley, D. A. 1994. *The President's Scientists: Reminiscences of a White House Science Advisor*. Yale University Press.

Bronowski, J. 1965. *Science and Human Values*. Harper & Row.

Brooks, H. 1964. The scientific adviser. In *Scientists and National Policy-Making*, ed. R. Gilpin and C. Wright. Columbia University Press.

Brooks, H. 1966. Basic science and agency mission. In *Research in the Service of National Purpose*, ed. F. Weyl. Government Printing Office.

Brooks, H. 1973. The physical sciences: Bellwether of science policy. In *Science and the Evolution of Public Policy*, ed. J. Shannon. Rockefeller University Press.

Brooks, H. 1980. Basic and applied research. In *Categories of Scientific Research*. National Science Foundation.

Brooks, H. 1985. The changing structure of the US research system. In H. Brooks and R. Schmitt, Current Science and Technology Issues: Two Perspectives, Occasional Paper No. 1, Graduate Program in Science, Technology, and Public Policy, George Washington University.

Brooks, H. 1994. The relationship between science and technology. *Research Policy* 23: 477–486.

Bruce-Briggs, B., ed. 1981. *The New Class?* McGraw-Hill.

Bundy, M. 1988. *Danger and Survival: Choices about the Bomb in the First Fifty Years*. Random House.

Burger, E. J., Jr. 1980. *Science at the White House: A Political Liability*. Johns Hopkins University Press.

Bush, V. 1945. *Science, the Endless Frontier*. Government Printing Office (reprinted in 1960).

Bush, V. 1950. Report of the Panel on the Gordon McKay Bequest Chaired by Vannevar Bush to the President and Fellows of Harvard College.

Carson, R. 1962. *Silent Spring*. Houghton Mifflin.

Casper, B. M., and Krauss, L. M. 1981. Fortune favors the prepared mind—a movement against nuclear war. *Science, Technology, & Human Values* 7, fall: 20–26.

Chayes, A., and Wiesner, J. B., eds. 1969. *ABM: An Evaluation of the Decision to Deploy an Antiballistic Missile System*. Harper & Row.

Clinton, W. J., and Gore, A. 1994. Science in the National Interest. Executive Office of the President, Office of Science and Technology Policy.

Cohen, I. B. 1995. *Science and the Founding Fathers: Science in the Political Thought of Jefferson, Franklin, Adams, and Madison*. Norton.

Collins, H. M. 1981. Stages in the empirical programme of relativism. *Social Studies of Science* 11: 3–10.

Committee for Economic Development—Research and Policy Committee. 1998. America's Basic Research: Prosperity Through Discovery. Committee for Economic Development.

Comroe, J. H., and Dripps, R. D. 1978. The Top Ten Medical Advances in Cardiovascular-Pulmonary Medicine and Surgery between 1945 and 1975, How They Came About: Final Report. Department of Health, Education, and Welfare, Public Health Service, National Institutes of Health.

Crow, M., and Bozeman, B. 1998. *Limited by Design: R&D Laboratories in the US National Innovation System*. Columbia University Press.

Culliton, B. J. 1979. Science's restive public. In *Limits of Scientific Inquiry*, ed. G. Holton and R. Morison. Norton.

Doble, J. 1995. Public opinion about issues characterized by technological complexity and scientific uncertainty. *Public Understanding of Science* 4: 95–118.

Dupree, A. H. 1986. *Science in the Federal Government: A History of Policies and Activities*. Johns Hopkins University Press.

Ehrenreich, H. 1997. Halbleiterforschung in den 50er Jahren: Anwendungsorientierte Forschung und Zukunftstechnologie. *Physikalische Blätter* 53: 21–26.

Evans, S. 1980. *Personal Politics: The Roots of Women's Liberation in the Civil Rights Movement and the New Left*. Vintage Books.

Ezrahi, Y. 1990. *The Descent of Icarus: Science and the Transformation of Contemporary Democracy*. Harvard University Press.

Fee, E. 1982. A feminist critique of scientific objectivity. *Science for the People* 14, no. 4: 5–8, 30–33.

Filner, R. E. 1976. The roots of political activism in British science. *Bulletin of the Atomic Scientists*, January: 25–29.

Fölsing, A. 1993. *Albert Einstein: A Biography*. Penguin.

Freeman, J. 1975. *The Politics of Women's Liberation: A Case Study of an Emerging Social Movement and Its Relation to the Policy Process*. Longman.

Fuqua, D. 1993. Federal investment in science and technology: Priorities for tomorrow. In *Science and Technology Advice to the President, Congress, and Judiciary*, ed. W. Golden. AAAS Press.

Galileo. 1957. *Discoveries and Opinions of Galileo*. Doubleday Anchor.

Gamson, W. A. 1968. *Power and Discontent*. Dorsey.

Gamson, W. A. 1975. *The Strategy of Social Protest*. Dorsey.

Gamson, W. A. 1992. The social psychology of collective action. In *Frontiers in Social Movement Theory*, ed. A. Morris and C. Mueller. Yale University Press.

Gamson, W. A. 1995. Hiroshima, the Holocaust, and the politics of exclusion. *American Sociological Review* 60: 1–20.

Garfinkel, M. S., and Weiss, S. C. 1999. In the court of history, Ehlers v. Bush. *Recent Science Letter* 1, spring: 6–7.

Gerlach, L. P., and Hine, V. H. 1970. *People, Power, Change: Movements of Social Transformation*. Bobbs-Merrill.

Gibbons, J. H. 1997. *This Gifted Age: Science and Technology at the Millennium*. AIP Press.

Gieryn, T. F. 1983. Boundary-work and the demarcation of science from non-science: Strains and interests in professional ideologies of scientists. *American Sociological Review* 48: 781–795.

Gieryn, T. F. 1999. *Cultural Boundaries of Science: Credibility on the Line*. University of Chicago Press.

Gilpin, R., and Wright, C., eds. 1964. *Scientists and National Policy-Making*. Columbia University Press.

Goertzel, T. G., and Goertzel, B. 1995. *Linus Pauling: A Life in Science and Politics*. Basic Books.

Golden, W. T., ed. 1993. *Science and Technology Advice to the President, Congress, and Judiciary*. AAAS Press.

Gottfried, K. 1999. Physicists in politics. *Physics Today* 52, March 3: 42–48.

Graham, J. D., ed. 1991. *Harnessing Science for Environmental Regulation*. Praeger.

Greenberg, D. S. 1966. "Hindsight": DOD study examines return on investment in research. *Science* 155, November 18: 872–873.

Greenwood, T. 1986. Science and technology advice for the president. In *The Presidency and Science Advising*, volume 1, ed. K. Thompson. University Press of America.

Griliches, Z. 1998. *R&D and Productivity: The Econometric Evidence*. University of Chicago Press.

Guston, D. H. 2000a. *Between Politics and Science: Assuring the Integrity and Productivity of Research*. Cambridge University Press.

Guston, D. H. 2000b. Retiring the Social Contract for Science. *Issues in Science and Technology* 16, no. 4: 32–36.

Habermas, J. 1975. *Legitimation Crisis*. Beacon.

Habermas, J. 1981. *Theorie des kommunikativen Handelns*. Suhrkamp.

Haines, H. H. 1984. Black radicalization and the funding of civil rights: 1957–1970. *Social Problems* 32: 31–43.

Haines, H. H. 1988. *Black Radicals and the Civil Rights Mainstream, 1954–1970*. University of Tennessee Press.

Halpern, J. 1997. The US National Academy of Sciences—In service to science and society. *Proceedings of the National Academy of Sciences* 94: 1606–1608.

Hardy, G. H. 1967. *A Mathematician's Apology*. Cambridge University Press. [first publ. 1940]

Hart, D. M. 1998. *Forged Consensus: Science, Technology, and Economic Policy in the United States, 1921–1953*. Princeton University Press.

Hedal, L. 1980. Citizen's League Against the Sonic Boom vs. the National Academy of Sciences: Scientific activists versus scientific advisors. A.B. thesis, Harvard University.

Herken, G. 1992. *Cardinal Choices: Presidential Science Advising from the Atomic Bomb to SDI*. Oxford University Press.

Holton, G. 1986. *The Advancement of Science, and Its Burdens*. Cambridge University Press. [second edition: Harvard University Press, 1998].

Holton, G. 1988. Jefferson, science, and national destiny. In *America in Theory*, ed. L. Berlowitz et al. Oxford University Press.

Holton, G. 1993. *Science and Anti-Science*. Harvard University Press.

Holton, G. 1998. What kinds of science are worth supporting? A new look, and a new mode. In *The Great Ideas Today*. Encyclopaedia Britannica.

Holton, G. 1999. I. I. Rabi as educator and science warrior. *Physics Today*, September: 37–42.

Holton, G. 2001. What is the imperative for basic science that serves national needs? In *Science for Society*, ed. L. Branscomb et al. Belfer Center for Science and International Affairs.

Holton, G., Chang, H., and Jurkowitz, E. 1996. How a scientific discovery is made: A case history. *American Scientist* 84: 364–375.

Holton, G., and Sonnert, G. 1999. A vision of Jeffersonian science. *Issues in Science and Technology* 16, no. 1: 61–65.

House Committee on Science. 1998. Unlocking Our Future: Toward a New National Science Policy. Report to Congress. (This is widely known as the Ehlers Report.)

Hubbard, R. 1986. Facts and feminism: Thoughts on the masculinity of natural science. *Science for the People* 18, no. 2: 16–20, 26.

Illinois Institute of Technology Research Institute 1968. *Technology in Retrospect and Critical Events in Science*. NSF contract NSF-C535. IIT Research Institute.

Inglehart, R. 1977. *The Silent Revolution: Changing Values and Political Styles Among Western Publics*. Princeton University Press.

Jacobs, P., and Landau, S. 1966. *The New Radicals: A Report with Documents*. Vintage Books.

Jasanoff, S. 1990. *The Fifth Branch: Science Advisers as Policymakers*. Harvard University Press.

Kegan, R. 1982. *The Evolving Self: Problem and Process in Human Development*. Harvard University Press.

Kevles, D. J. 1987. *The Physicists: The History of a Scientific Community in Modern America*. Harvard University Press.

Keyworth, G. A., II 1993. Science advice during the Reagan years. In *Science and Technology Advice to the President, Congress, and Judiciary*, ed. W. Golden. AAAS Press.

Killian, J. R., Jr. 1977. *Sputnik, Scientists, and Eisenhower: A Memoir of the First Special Assistant to the President for Science and Technology*. MIT Press.

Kistiakowsky, G. B. 1976. *A Scientist at the White House: The Private Diary of President Eisenhower's Special Assistant for Science and Technology*. Harvard University Press.

Klages, L. 1956. Mensch und Erde (originally published in 1913). In Klages, L., *Mensch und Erde: Zehn Abhandlungen*. Alfred Kröner.

Klapp, O. E. 1972. *Currents of Unrest: An Introduction to Collective Behavior*. Holt, Rinehart and Winston.

Koertge, N., ed. 1998. *A House Built on Sand: Exposing Postmodernist Myths about Science*. Oxford University Press.

Kohlberg, L. 1984. *The Psychology of Moral Development: The Nature and Validity of Moral Stages*. Harper & Row.

Kreckel, R., et al. 1986. *Regionalistische Bewegungen in Westeuropa: Zum Struktur- und Wertwandel in fortgeschrittenen Industriegesellschaften*. Leske.

Kuznick, P. J. 1987. *Beyond the Laboratory: Scientists as Political Activists in 1930s America*. University of Chicago Press.

Lam, M. S. 1994. Women and Men Scientists' Notions of the Good Life: A Developmental Approach. Ph.D. dissertation, University of Massachusetts, Amherst.

Lanouette, W. 1992. *Genius in the Shadows: A Biography of Leo Szilard, the Man Behind the Bomb*. Scribner.

Latour, B. 1987. *Science in Action: How to Follow Scientists and Engineers through Society*. Harvard University Press.

Latour, B., and Woolgar, S. 1979. *Laboratory Life: The Social Construction of Scientific Facts*. Sage.

Leslie, S. W. 1993. *The Cold War and American Science: The Military-Industrial-Academic Complex at MIT and Stanford*. Columbia University Press.

Lipset, S. M. 1950. *Agrarian Socialism: The Cooperative Commonwealth Federation in Saskatchewan*. University of California Press.

Lo, C. Y. H. 1992. Communities of challengers in social movement theory. In *Frontiers in Social Movement Theory*, ed. A. Morris and C. Mueller. Yale University Press.

Lowen, R. S. 1997. *Creating the Cold War University: The Transformation of Stanford*. University of California Press.

Loevinger, J., with A. Blasi. 1976. *Ego Development*. Jossey-Bass.

Lofland, J. 1989. Consensus movements: City twinning and derailed dissent in the American eighties. *Research in Social Movements, Conflicts and Change* 11: 163–196.

Luhmann, N. 1990. *Die Wissenschaft der Gesellschaft*. Suhrkamp.

Marshall, E. 1980. Frank Press's number game. *Science* 210, October 24: 406.

Mazuzan, G. T. 1994. The National Science Foundation: A Brief History. NSF document 88–16.

McAdam, D. 1982. *Political Process and the Development of Black Insurgency, 1930–1970*. University of Chicago Press.

McAdam, D. 1983. Tactical innovation and the pace of insurgency. *American Sociological Review* 48: 735–754.

McAdam, D. 1986. Recruitment to high-risk activism: The case of Freedom Summer. *American Journal of Sociology* 92: 64–90.

McAdam, D., McCarthy, J. D., and Zald, M. N. 1988. Social movements. In *Handbook of Sociology*, ed. N. Smelser. Sage.

McAdam, D., McCarthy, J. D., and Zald, M. N. 1996a. Introduction: Opportunities, mobilizing structures, and framing processes—toward a synthetic, comparative perspective on social movements. In *Comparative Perspectives on Social Movements*, ed. D. McAdam et al. Cambridge University Press.

McAdam, D., McCarthy, J. D., and Zald, M. N., eds. 1996b. *Comparative Perspectives on Social Movements: Political Opportunities, Mobilizing Structures, and Cultural Framings*. Cambridge University Press.

McCarthy, J. D. 1987. Pro-life and pro-choice mobilization: Infrastructure deficits and new technologies. In *Social Movements in an Organizational State*, ed. M. Zald and J. McCarthy. Transaction Books.

McCarthy, J. D., and Wolfson, M. 1992. Consensus movements, conflict movements, and the cooptation of civic and state infrastructures. In *Frontiers in Social Movement Theory*, ed. A. Morris and C. Mueller. Yale University Press.

McCarthy, J. D., and Zald, M. N. 1973. *The Trend of Social Movements in America: Professionalization and Resource Mobilization*. General Learning Press.

McCarthy, J. D., and Zald, M. N. 1977. Resource mobilization and social movements: A partial theory. *American Journal of Sociology* 82: 1212–1241.

McPhail, C. 1971. Civil disorder participation: A critical examination of recent research. *American Sociological Review* 36: 1058–1073.

Merton, R. K. 1968. *Social Theory and Social Structure*. Free Press.

Merton, R. K. 1973. *The Sociology of Science: Theoretical and Empirical Investigations*. University of Chicago Press.

Mervis, J. 2001. NSF scores low on using own criteria. *Science* 291, 30 March: 2533–2535.

Michels, R. 1949. *Political Parties: A Sociological Study of the Oligarchical Tendencies of Modern Deomcracy*. Free Press (first published in 1911).

Miller, J. D. 1999. Scientific literacy, issue attentiveness, and attitudes toward science and space exploration. Paper presented to American Astronomical Society, Austin, Texas.

Miller, J. D. 2000. Public acquisition and retention of scientific information: Recent patterns and future strategies. Paper presented to National Science Board Symposium on Communicating Science and Technology in the Public Interest, Irvine, California.

Mitchell, R. C. 1979. National environmental lobbies and the apparent illogic of collective action. In *Collective Decision-Making: Applications from Public Choice Theory*, ed. C. Russell. Johns Hopkins University Press.

Mitroff, I. I. 1974. Norms and counter-norms in a select group of Apollo moon scientists: A case study of the ambivalence of scientists. *American Sociological Review* 39: 579–595.

Moore, K. 1993. Doing Good While Doing Science: The Origins and Consequences of Public Interest Science Organizations in America, 1945–1990. Dissertation, University of Arizona.

Moore, K. 1996. Organizing integrity: American science and the creation of public interest organizations, 1955–1975. *American Journal of Sociology* 101: 1592–1627.

Morison, R. S. 1979. Introduction. In *Limits of Scientific Inquiry*, ed. G. Holton and R. Morison. Norton.

Morris, A. D., and Mueller, C. M., eds. 1992. *Frontiers in Social Movement Theory*. Yale University Press.

Morris, A. D. 1984. *The Origins of the Civil Rights Movement: Black Communities Organizing for Change*. Free Press.

Mowery, D. C., and Rosenberg, N. 1998. *Paths of Innovation: Technological Change in 20th-Century America*. Cambridge University Press.

Mullins, N. C. 1981. Power, social structure, and advice in American science: The United States national advisory system, 1950–1972. *Science, Technology, & Human Values* 7, fall: 4–19.

NAS (National Academy of Sciences). 1964. Federal Support of Basic Research in Institutions of Higher Learning. National Academy of Sciences and National Research Council.

NAS. 1965. Basic Research and National Goals. A Report to the Committee on Science and Astronautics, US House of Representatives. Government Printing Office.

NAS. 1967. Applied Science and Technological Progress. A Report to the Committee on Science and Astronautics, US House of Representatives. Government Printing Office.

NAS. 1969. Technology: Processes of Assessment and Choice. A Report to the Committee on Science and Astronautics, US House of Representatives. Government Printing Office.

Nelkin, D. 1971. *Nuclear Power and its Critics: The Cayuga Lake Controversy*. Cornell University Press.

Nelkin, D. 1972. *The University and Military Research: Moral Politics at MIT*. Cornell University Press.

Nelkin, D. 1979. Threats and promises: Negotiating the control of research. In *Limits of Scientific Inquiry*, ed. G. Holton and R. Morison. Norton.

Nelkin, D., ed. 1984a. *Controversy: Politics of Technical Decisions*. Sage.

Nelkin, D. 1984b. Science, technology, and political conflict: Analyzing the issues. In *Controversy: Politics of Technical Decisions*, ed. D. Nelkin. Sage.

Nelkin, D. 1987. *Selling Science: How the Press Covers Science and Technology*. Freeman.

Nelson, R. R., and Phelps, E. S. 1965. *Investment in Humans, Technological Diffusion and Economic Growth*. RAND.

Nelson, R. R. 1996. *The Sources of Economic Growth*. Harvard University Press.

Nichols, D. 1974. The associational interest groups of American science. In *Scientists and Public Affairs*, ed. A. Teich. MIT Press.

Nichols, R. W. 1986. Pluralism in science and technology: Arguments for organizing federal support for R&D around independent missions. *Technology in Society* 8: 33–63.

NSB (National Science Board). 1978. *Basic Research in the Mission Agencies: Agency Perspectives on the Conduct and Support of Basic Research*. Government Printing Office.

NSB. 1998. *Science & Engineering Indicators—1998*. National Science Foundation.

NSB. 2000. *Science & Engineering Indicators—2000*. National Science Foundation.

NSF (National Science Foundation). 1976. Science for Citizens: A Program Plan (draft copy). National Science Foundation

NSF. 1980a. *How Basic Research Reaps Unexpected Rewards*. National Science Foundation.

NSF. 1980b. *The Five-Year Outlook: Problems, Opportunities and Constraints in Science and Technology*. Government Printing Office.

Olson, M. 1965. *The Logic of Collective Action: Public Goods and the Theory of Groups*. Harvard University Press.

Omenn, G. S. 1984. Basic research as an investment in the nation's future. In *The Impact of Protein Chemistry on the Biomedical Sciences*, ed. A. Schechter et al. Academic Press.

Omenn, G. S. 1985. Basic research as an investment in the future. *American Journal of Roentgenology* 145: 1109–1112.

Orum, A. M. 1972. *Black Students in Protest: A Study of the Origins of the Black Student Movement*. American Sociological Association.

Pauling, L., and Teller, E. 1958. Fallout and disarmament: A debate between Linus Pauling and Edward Teller. *Daedalus* 87, spring: 147–163.

Perl, M. L. 1971. The scientific advisory system: Some observations. *Science* 173: 1211–1215.

Perrow, C. 1970. *Organizational Analysis: A Sociological View*. Wadsworth.

Pinard, M. 1975. *The Rise of a Third Party: A Study in Crisis Politics*. McGill–Queen's University Press.

Press, F. 1978a. Science and technology: The road ahead. *Science* 200, 19 May: 737–741.

Press, F. 1978b. An agenda for technology and policy. *Technology Review* 80: 51–55.

Press, F. 1981a. Science and technology in the White House, 1977 to 1980: Part 1. *Science* 211, January 9: 139–145.

Press, F. 1981b. Science and technology in the White House, 1977 to 1980: Part 2. *Science* 211, January 16: 249–256.

Press, F. 1993. A prescription for science and technology advising for the President: An interview by William T. Golden. In *Science and Technology Advice to the President, Congress, and Judiciary*, ed. W. Golden. AAAS Press.

Press, F. 1995. Growing up in the Golden Age of science. *Annual Review of Earth and Planetary Sciences* 23: 1–9.

Price, D. J. de S. 1961. *Science since Babylon*. Yale University Press.

Price, D. J. de S. 1984. Of sealing wax and string. *Natural History* 93, no. 1: 49–56.

Price, D. K. 1985. *America's Unwritten Constitution: Science, Religion, and Political Responsibility*. Harvard University Press.

Price, D. K. 1979. Endless frontier or bureaucratic morass? In *Limits of Scientific Inquiry*, ed. G. Holton and R. Morison. Norton.

Price, D. K. 1965. *The Scientific Estate*. Belknap.

Price, D. K. 1962. *Government and Science: Their Dynamic Relation in American Democracy*. Oxford University Press.

Primack, J., and von Hippel, F. 1974. *Advice and Dissent: Scientists in the Political Arena*. Basic Books.

PSAC (President's Science Advisory Committee). 1960. Scientific Progress, the Universities, and the Federal Government.

Rhodes, R. 1986. *The Making of the Atomic Bomb*. Simon & Schuster.

Rigden, J. S. 2000. *Rabi: Scientist and Citizen*. Harvard University Press.

Ringer, F. K. 1990. *The Decline of the German Mandarins: The German Academic Community, 1890–1933*. University Press of New England.

Rosenberg, N. 1972. *Technology and American Economic Growth*. Harper & Row.

Rotblat, J. 1999. A Hippocratic Oath for scientists. *Science* 286, no. 19: 1475.

Rothman, S., and Lichter, S. R. 1982. The nuclear energy debate: Scientists, the media and the public. *Public Opinion* 5, no. 4: 47–52.

Rupp, L., and Taylor, V. 1987. *Survival in the Doldrums: The American Women's Rights Movement, 1945 to the 1960s*. Oxford University Press.

Russell, B. 1960. The social responsibilities of scientists. *Science* 131, February 12: 391–392.

Sainte-Beuve, C. A. 1837. *Pensées d'août, Poésies*. Eugène Renduel.

Salomon, J.-J. 1973. Science and scientists' responsibilities in today's society. In *Scientists in Search of Their Conscience*, ed. A. Michaelis and H. Harvey. Springer-Verlag.

Salomon, J.-J. 2000. Science, technology and democracy. *Minerva* 38: 33–51.

Salomon, J.-J. 2001. Science policies in a new setting. *International Social Science Journal* 53: 323–335.

Schwartz, M., and Paul, S. 1992. Resource mobilization versus the mobilization of people: Why consensus movements cannot be instruments of social change. In *Frontiers in Social Movement Theory*, ed. A. Morris and C. Mueller. Yale University Press.

Segerstråle, U., ed. 2000. *Beyond the Science Wars: The Missing Discourse about Science and Society*. State University of New York Press.

Seitz, F. 1994. *On the Frontier: My Life in Science*. American Institute of Physics.

Shadish, W. R. 1989. The perception and evaluation of quality in science. In *The Psychology of Science. Contributions to Metascience*, ed. B. Gholson et al. Cambridge University Press.

Sherwin, C. W., and Isenson, R. S. 1967. Project Hindsight: A Defense Department study of the utility of research. *Science* 156, June 23: 1571–1577.

Shorter, E., and Tilly, C. 1974. *Strikes in France, 1830–1968*. Cambridge University Press.

Shurcliff, W. A. 1970. *S/S/T and Sonic Boom Handbook*. Ballantine.

Sills, D. L. 1957. *The Volunteers: Means and Ends in a National Organization*. Free Press.

Skocpol, T. 1997. The Tocqueville problem: Civic engagement in American democracy. *Social Science History* 21: 455–479.

Smith, A. K. 1965. *A Peril and a Hope: The Scientists' Movement in America: 1945–47*. University of Chicago Press.

Smith, B. L. R. 1992. *The Advisers: Scientists in the Policy Process*. Brookings Institution.

Snow, D. A., Rochford, E. B., Worden, S. K., and Benford, R. D. 1986. Frame alignment processes, micromobilization, and movement participation. *American Sociological Review* 51: 464–481.

Society for Freedom in Science 1953. *The Society for Freedom in Science: Its Origins, Objects and Constitution*. Society for Freedom in Science.

Solow, R. 1957. Technical change and aggregate production function. *Review of Economics and Statistics* 39: 312–320.

Sonnert, G. 1987. *Nationalismus und Krise der Moderne: Theoretische Argumentation und empirische Analyse am Beispiel des neueren schottischen Nationalismus*. Athenaeum.

Sonnert, G. 1995. What makes a good scientist?: Determinants of peer evaluation among biologists. *Social Studies of Science* 25: 35–55.

Sonnert, G., with G. Holton. 1995a. *Gender Differences in Science Careers: The Project Access Study*. ASA Rose Book Series. Rutgers University Press.

Sonnert, G., with G. Holton. 1995b. *Who Succeeds in Science? The Gender Dimension*. Rutgers University Press.

Sonnert, G., and Brooks, H. 2001. The basic-applied dichotomy in science policy: Lessons from the past. In *Science for Society*, ed. L. Branscomb et al. Belfer Center For Science and International Affairs.

Standish, L. 1979. Women, work, & the scientific enterprise: Solitary science vs. connected collectivism. *Science for the People* 11, no. 5: 12–18.

Steelman, J. R. 1947. *Science and Public Policy*. Government Printing Office.

Stokes, D. E. 1997. *Pasteur's Quadrant: Basic Science and Technological Innovation*. Brookings Institution.

Strickland, D. A. 1968. Scientists in Politics: The Atomic Scientists Movement, 1945–46. Purdue Research Foundation.

Strickland, S. P. 1989. *The Story of the NIH Grants Programs*. University Press of America.

Szilard, L. 1978. *Leo Szilard, His Version of the Facts: Selected Recollections and Correspondence*, ed. S. Weart and G. Szilard. MIT Press.

Tarrow, S. 1994. *Power in Movement: Social Movements, Collective Action and Politics*. Cambridge University Press.

Thompson, K. W., ed. 1986–1994. *The Presidency and Science Advising*. University Press of America.

Tilly, C. 1978. *From Mobilization to Revolution*. Addison-Wesley.

Tocqueville, A. de. 1862. *Democracy in America*, volume 2. Sever and Francis (originally published in France in 1840).

Toulmin, S. E. 1975. The twin moralities of science. In *Science and Society*, ed. N. Steneck. University of Michigan Press.

Trenn, T. J. 1983. *America's Golden Bough: The Science Advisory Intertwist*. Oelgeschlager, Gunn & Hain.

Useem, B. 1980. Solidarity model, breakdown model, and the Boston anti-busing movement. *American Sociological Review* 45: 357–369.

von Hippel, F. 1991. *Citizen Scientist*. AIP Press.

Walsh, E. J. 1981. Resource mobilization and citizen protest in communities around Three Mile Island. *Social Problems* 29: 1–21.

Walsh, E. J., and Warland, R. H. 1983. Social movement involvement in the wake of a nuclear accident: Activists and free riders in the TMI area. *American Sociological Review* 48: 764–780.

Wang, J. 1999. *American Science in an Age of Anxiety: Scientists, Anticommunism, and the Cold War*. University of North Carolina Press.

Weart, S. R. 1979. *Scientists in Power*. Harvard University Press.

Weart, S. R. 1988. *Nuclear Fear: A History of Images*. Harvard University Press.

Weiner, C. 1979. The recombinant DNA controversy: Archival and oral history resources. *Science, Technology, and Human Values* 26: 17–19.

Wiesner, J. B. 1993. The rise and fall of the President's Science Advisory Committee. In *Science and Technology Advice to the President, Congress, and Judiciary*, ed. W. Golden. AAAS Press.

Wilson, K. L., and Orum, A. M. 1976. Mobilizing people for collective political action. *Journal of Political and Military Sociology* 4: 187–202.

Wise, G. 1985. Science and technology. *Osiris* 1, second series: 229–246.

Zachary, G. P. 1997. *Endless Frontier: Vannevar Bush, Engineer of the American Century*. Free Press.

Zald, M. N., and McCarthy, J. D. 1975. Organizational intellectuals and the criticism of society. *Social Service Review* 49: 344–362.

Zuckerman, S. 1971. *Beyond the Ivory Tower: The Frontiers of Public and Private Science*. Taplinger.

Index

Made in the USA
Monee, IL
05 November 2026

67811376R00134